Problem Solving

for

Physics
A World View

Fifth Edition

Larry Kirkpatrick
Montana State University

Gregory E. Francis
Montana State University

Australia • Canada • Mexico • Singapore • Spain • United Kingdom • United States

Printed in the United States of America
2 3 4 5 6 7 07 06 05 04

Printer: Globus Printing & Packaging
Image credit: Warren Bolster/Getty Images

0-534-40825-7

For more information about our products, contact us at:
Thomson Learning Academic Resource Center
1-800-423-0563

For permission to use material from this text, contact us by:
Phone: 1-800-730-2214
Fax: 1-800-730-2215
Web: http://www.thomsonrights.com

Asia
Thomson Learning
5 Shenton Way #01-01
UIC Building
Singapore 068808

Australia/New Zealand
Thomson Learning
102 Dodds Street
Southbank, Victoria 3006
Australia

Canada
Nelson
1120 Birchmount Road
Toronto, Ontario M1K 5G4
Canada

Europe/Middle East/South Africa
Thomson Learning
High Holborn House
50/51 Bedford Row
London WC1R 4LR
United Kingdom

Latin America
Thomson Learning
Seneca, 53
Colonia Polanco
11560 Mexico D.F.
Mexico

Spain/Portugal
Paraninfo
Calle/Magallanes, 25
28015 Madrid, Spain

PREFACE

Welcome

We have written *Problem Solving to Accompany PHYSICS: A World View* to develop some of the numerical aspects of your study of physics that can be accessed with simple algebra and geometry. The only course that you need to have taken is the first year algebra course offered in most high schools.

PHYSICS: A World View is intended as a conceptual view of physics. Many discussions in the text are presented without the underlying mathematics. This supplement takes you one step further into the mathematics that provides the foundation for the physics world view. This manual cannot be used by itself. It assumes you are reading the primary text, *PHYSICS: A World View*, and it relies on the physics covered there.

You will see many sections in the textbook that have a boxed ✓MATH printed to the right of the section heading. This icon means that some of the ideas discussed in the section are developed further in this supplement. Read the primary text before turning to this manual. The supplement starts where the primary text ends; it makes little attempt to go back over the material.

Why Use Numbers?

In many ways mathematics is the language of physics. While the basic concepts of our world view often originate in our physical descriptions and thus can be expressed in words, it is usually the mathematical manipulations of these concepts that give us powerful insights. In our development of a physics world view, we will see many instances where such mathematical manipulations have generated new physical concepts.

Just as you don't have to have the talent to paint a masterpiece in order to appreciate a great painting, you do not need to be mathematically skilled in order to appreciate the physics world view. However, it often helps to do a few simple calculations while developing an understanding of a basic physical concept. It is in this spirit that we have written this mathematical supplement.

Organization

This manual begins with an introductory chapter on significant digits and units. The rest of the chapters follow the organization of the text. The topics are mostly independent of each other and need not be studied in order.

The sections of this manual contain numerous worked examples and practice problems. The answers to the practice problems are placed immediately following for easy reference. At the end of each chapter, there are additional problems. The more challenging problems are marked with an asterisk. Answers to the odd-numbered problems are given at the end of the manual. Solutions to all the problems are provided in the *Instructor's Resource Manual*.

Appendix

We have collected the values for many of the commonly used physical constants and physical data in an appendix at the end of the manual. This appendix also includes a list of standard abbreviations, prefixes for the powers-of-ten notation, conversion factors between different units, the Greek alphabet, and useful data for the Solar System. To solve some of the problems you will need to refer to the appendix to get numerical data.

We hope that the material in this mathematical supplement will complement the conceptual ideas that you will learn in the textbook. The combination of the two views will give you a more complete understanding of physicists' view of the world around us.

Larry D. Kirkpatrick
Gregory E. Francis

Montana State University

April 2003

PROBLEM SOLVING

to Accompany

PHYSICS: A WORLD VIEW

Table of Contents

Table of Contents

1—A WORLD VIEW

1.1 How Many Digits?

When you divide 73 by 13 on your calculator, the display reads 5.6153846. But how do you know how many digits to record in your answer? In mathematics, there may be no reason to worry about it. In science, however, the numbers 73, 13, and the answer usually represent physical quantities. If we keep too many digits in the answer, we imply a greater precision in the calculated quantity than in the measured quantities. The digits that reflect the precision of the measurements are called *significant digits*.

Although there is a set of rules that can tell you how many digits to keep in real-world situations, we will not need to learn these rules. We will simply agree to keep no more than three significant digits in our answers. Therefore,

$$\frac{73}{13} = 5.62$$

In effect we are assuming that the given data have three significant digits even if they are not written out explicitly. In this case, we are assuming that 73.0/13.0 = 5.62. In this same spirit, we will usually not keep trailing zeros after the decimal point in our answers.

When keeping the three significant digits, we will round our answer before dropping the extra digits. To do this, look at the fourth digit from the left. If it is less than 5, drop the remaining digits. If it is 5 or greater, add 1 to the third digit and then drop the fourth and any remaining digits. The following examples illustrate this procedure:

Example 1.1.1

Round the numbers 12.343 21 and 2.345 678 9 to three significant digits.

Look at the fourth digit from the left in the first number. Because this 4 is less than 5, we simply drop it and the remaining digits.

$$12.343\ 21 \rightarrow 12.3$$

In the second number, the fourth digit is a 5. Therefore, we add one to the third digit before dropping the remaining digits.

$$2.345\ 678\ 9 \rightarrow 2.35$$

Practice: Round the following numbers to three significant digits: (a) 6.783 (b) 9.846 (c) 34.89 (d) 56.34
Answers: 6.78; 9.85; 34.9; and 56.3

If the number is smaller than 1, we ignore the leading zeros in counting significant digits because their only purpose is to locate the decimal point. Therefore, 0.013 has only two significant digits. Similarly, if the number is larger than 1000, such as 15,325, we replace the "dropped" digits with zeros to keep the decimal in its proper place. Therefore, we would round this number to read 15,300.

Example 1.1.2

Round the following numbers to three significant digits: 0.003 412 and 19,651.

In the first number we look for the first non-zero digit and count over three more digits. Because this 2 is less than 5, we drop it and all following digits.

$$0.003\,412 \rightarrow 0.003\,41$$

In the second number we add one to the third digit, because the fourth digit is a 5. Then we convert all of the digits beyond the 7 to zeros.

$$19{,}651 \rightarrow 19{,}700$$

Practice: Round the following to three significant digits: (a) 0.028 632 (b) 31,557,082
Answers: 0.0286; 31,600,000

You may keep additional digits in your calculator during computations and round off the answer at the end. If you are doing the calculations by hand, you can save yourself work by rounding off intermediate results to 4 digits. The small differences in the answers are not important.

In the text, we rounded the value of the acceleration due to gravity from 9.8 m/s^2 to 10 m/s^2 to make calculations easier. In this manual, we will use three significant digits as we do with most other numbers.

There are occasions where we will use more than three significant digits. This usually happens when an effect is hidden when only three significant digits are used. As an example, in Chapter 26 we calculate the mass difference between a nucleus and its constituent protons and neutrons. The mass difference typically occurs in the third significant digit and would not be very accurate unless we use additional digits.

1.2 Units

It is tempting to ignore the units associated with numerical values, but this can be the source of trouble. For instance, if you ask for a board with a length of 8, it is not clear if you need a board that is 8 inches, 8 feet, or 8 yards long. Obviously, you will be upset if you ordered an 8-ft stud and got an 8-in. stick. In communicating your numerical results, it is very important to state your units.

You have probably been told many times that you cannot add apples and oranges. This is a reminder that when you add or subtract numbers, they must have the same units. You obviously cannot add a length to a time, but you must also be careful when adding lengths. You must make sure that all of your length measurements are in the same units.

It is a very good habit to include the units for all numbers used in calculations. These units produce the units in the answer and help avoid problems. For instance, the units are a check that we have used a consistent procedure. If we incorrectly calculate a speed and get a number with units of length × time, we know that we have made an error because speed has units of length/time. Units are also a check that we have used a consistent set of measurements. We should not use the speed of a car in miles per hour in an equation that requires that all speeds be given in meters per second. Our answer will not make any sense.

The units can also serve as a check to be sure that we do not leave out one of the measurements in a calculation. We should always manipulate the units using the rules of algebra to see if they give the expected units for the answer. In some cases, we can even guess the form of the equation knowing the units of the answer.

Example 1.2.1

How far does a car travel if it has a constant speed of 80 km/h for 2.5 h?

$$d = vt = \left(80\frac{\text{km}}{\text{h}}\right)(2.5\text{ h})$$

Canceling the two h's, we see that our answer is 200 km. Suppose that during an exam you incorrectly divided the speed by the time. You would have obtained 32 km/h^2, something that is only seen as nonsense when you examine the units.

Practice: How far does an airplane travel in 3.2 h at a constant speed of 800 km/h?
Answer: 2560 km

1.3 Changing Units

There are often times when we need to convert a measurement in one set of units into another set of units. There shouldn't be any confusion about whether to multiply or divide by the conversion factor if you follow a definite procedure each time.

The procedure uses multiplication by 1. This works because 1 is the only number that does not change the value of the measurement. However, it can change the *form* of the number. The technique involves writing 1 as a fraction with its numerator equal to its denominator. For instance, we have learned that 1 foot is the same as 12 inches. Thus,

$$\frac{12\text{ in.}}{1\text{ ft}} = 1 \quad \text{or} \quad \frac{1\text{ ft}}{12\text{ in.}} = 1$$

We can now convert measurements in feet to the equivalent values in inches (or vice versa) by multiplying by 1 in one of these forms. How do you know which form to use? Always choose the fraction that cancels out the old unit. This will automatically put the new unit in its proper place.

Example 1.3.1

How many inches are there in 8 ft?

Because we want to cancel out the ft, we use the first form of 1.

$$8\text{ ft}\left[\frac{12\text{ in.}}{1\text{ ft}}\right] = 96\text{ in.}$$

Practice: How many feet are there in 100 inches?
Answer: 8.33 ft

You can do the conversions in several steps, such as converting miles to feet and then feet to inches by multiplying by both forms of 1. You can also convert two units at the same time. A list of conversion factors is given in the Appendix at the back of this manual that can be used to make the "1" fractions.

Example 1.3.2

What is a speed of 60 mph expressed in ft/s?

We know that 1 mile = 5280 ft, 1 h = 60 min, and 1 min = 60 s. Therefore,

$$\left(60\frac{\text{miles}}{\text{h}}\right)\left[\frac{5280\text{ ft}}{1\text{ mile}}\right]\left[\frac{1\text{ h}}{60\text{ min}}\right]\left[\frac{1\text{ min}}{60\text{ s}}\right]=88\frac{\text{ft}}{\text{s}}$$

Practice: If you make $4.80 per hour, how many cents per minute do you make?
Answer: 8¢/min

1.4 Problem Solving

How long does it take to pass a squeeze around the world? When you first read this question, you may ask yourself, "What does this mean?" This same question will come to mind when you read some of the physics questions in the text or this manual. The first step in solving a problem is to figure out what question is being asked.

In this case you can image a group of friends holding hands. When a friend squeezes one of your hands, you squeeze the hand of the friend on your other side. Now imagine a line of people stretching clear around Earth along the equator. As each person's hand is squeezed, the person squeezes the hand of the next person in line. It is like a "wave" moving around a football or soccer stadium.

Before we do the calculation, how long do you think it will take to pass the squeeze around the world? Will it take less than a day? Less than a week? Less than a month? Or, less than a year? It is a good idea to estimate your answer before doing a calculation. This will give you a check on your method. It is also a good idea to examine your answer to see if it is at least within the realm of possibilities.

In most physics problems you will be given a set of numbers to use in the calculations. At times you will need to look up a number or two in the text. In this particular problem we need to know the distance around Earth. It is approximately 25,000 miles, or 40,000 km = 40,000,000 m. In the spirit of "going metric," let's use the latter number.

In some physics problems you will need to estimate some of the values used in the problem. In this problem, we need to know how far apart the people are standing in order to calculate how many people we need to form the line around Earth. It would be very tiring to stand with your arms straight out, so let's assume that the arms are relaxed like they were walking down the sidewalk holding hands. Therefore, the people are roughly 1 m apart. The number of people is therefore equal to the number of meters, or 40,000,000 people.

We now need to estimate how long it takes each person, on the average, to pass the squeeze. If the people are not paying attention to what's going on, this could take a second or two. If they are watching the squeeze approach down the line, they could pass the squeeze along in less than a tenth of a second. However, if they are aware that their time is approaching, but are not watching the line, a typical reaction time is about 0.3 s. Let's assume this time for our calculation.

$$40{,}000{,}000 \text{ people}\left(\frac{0.3 \text{ s}}{\text{person}}\right) = 12{,}000{,}000 \text{ s}$$

This is our answer. But just how long is 12 million seconds? Our answer will make more sense if we convert it into longer time units. Let's convert our answer to days.

$$(12{,}000{,}000 \text{ s})\left[\frac{1 \text{ min}}{60 \text{ s}}\right]\left[\frac{1 \text{ h}}{60 \text{ min}}\right]\left[\frac{1 \text{ day}}{24 \text{ h}}\right] = 139 \text{ days}$$

Because we estimated the average spacing and the average reaction time, we are not justified in saying it will take 139 days. The time will be around 139 days, but it could be many days longer or shorter than this. Thus, it takes approximately four to five months to pass a squeeze around Earth. How well did you guess the result?

Problems

1. Round the following numbers to 3 significant digits: (a) 3.337 (b) 38,627 (c) 0.666 66 (d) 0.001 234 5
2. Round the following numbers to 3 significant digits: (a) 0.765 43 (b) 0.003 636 (c) 7.4141 (d) 55,567
3. Round the following answers to 3 significant digits: (a) 43/6 (b) 77×777
4. Round the following answers to 3 significant digits: (a) 3.56×47.9 (b) 0.82/0.242
5. Round the following answers to 3 significant digits: (a) $(5.82 \times 10^3)/(1.22 \times 10^6)$ (b) $(2.34 \times 10^{-3})(8.76 \times 10^4)$
6. Round the following answers to 3 significant digits: (a) $(3.873 \times 10^2)/(2.222 \times 10^5)$ (b) $(3.333 \times 10^5)(3.895 \times 10^2)$
7. An airplane flies 1670 miles between two cities in 3.28 h. What is the airplane's average speed? Don't forget to use the correct number of significant digits and the correct units.
8. A car travels 543 km in 4.52 h. What is the car's average speed? Don't forget to use the correct number of significant digits and the correct units.
9. What is the volume of a cube with edges that are 1.25 m long?
10. What is the area of a square with sides that are 3.21 m long?
11. How many inches are there in 1 mile?
12. If there are 16 oz in 1 lb, how many ounces are there in 1 ton (2000 lb)?
13. Given that 2.54 cm = 1 in., how tall is a 6-ft person in centimeters?
14. Given that there are 2.54 cm in 1 in., how many centimeters are there in 1 ft?
15. How many millimeters are there in 5.2 km?

16. How many seconds are there in a typical lifetime?
17. Given that 1 mile = 1.61 km, what is a speed of 70 mph expressed in km/h?
18. Given that 1 mile = 1.61 km, what is a speed of 120 km/h expressed in mph?
19. A car has a speed of 32 m/s. What is this speed expressed in km/h?
20. What is a speed limit of 120 km/h expressed in m/s?
21. How many square centimeters are there in an area of 1 m^2?

*22. How many cubic centimeters are there in a box with a volume of 1 m^3?

*23. How many breaths does an average person take in a lifetime?

*24. How days would it take to walk across the United States?

*25. Approximately how many gallons of gasoline are burned per day in cars in the U.S?

*26. Approximately how many tons of aluminum does it take to make the cans for soft drinks consumed per day in the U.S?

2 — DESCRIBING MOTION

2.1 Images of Speed

One way of representing motion was discovered when a French mathematician invented the technique of graphing. Although there are a number of types of graph, we will focus on the common two-axis graph that represents the relationship between two quantities. This type of graph displays relationships between a variety of quantities such as the rise and fall of the Dow Jones Industrial Average with time, your weight on each day of the year, or the distance it takes to stop a car traveling at different speeds.

We will use graphs to represent the motion of a puck and show how this type of representation is especially valuable for analyzing motion. The horizontal axis will represent time. For convenience, imagine that we start a stopwatch at the beginning of the motion. Therefore, the time axis usually starts with zero and advances into the future. The vertical axis represents the position of the object. Because time and position are totally separate quantities, there is no relationship between the two scales; we are free to choose each one independently to best represent the situation.

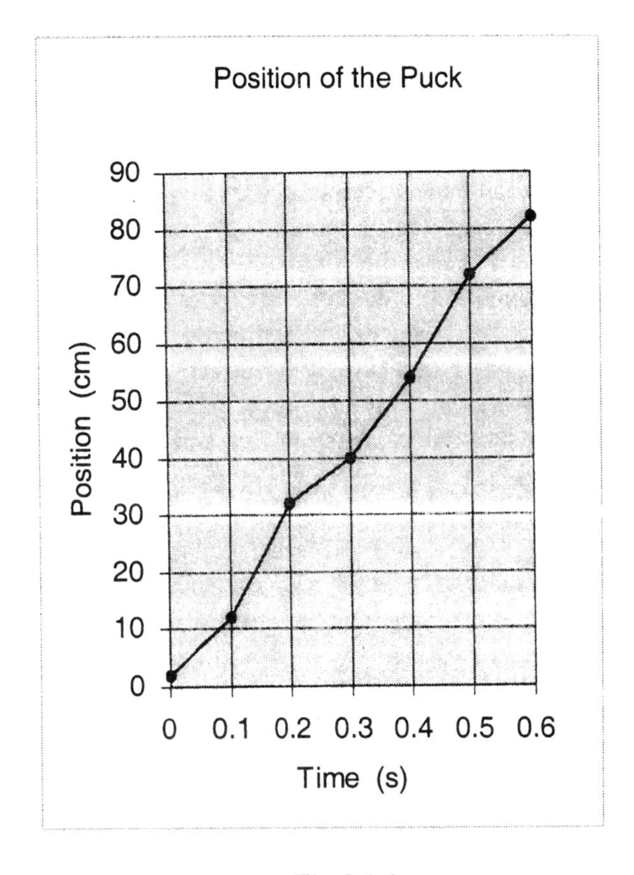

Fig 2.1.1

Let's assume that the positions of the puck are 2, 12, 32, 40, 54, 72, and 82 cm. Because the puck traveled between 2 cm and 82 cm, a good choice for the vertical axis is a scale ranging from 0 to 90 cm with divisions every 10 cm. The strobe images occurred every 0.1 second. Because there are 6 images, the scale for the time axis is chosen to range from 0 to 0.6 s with divisions each 0.1 s.

The graph in Figure 2.1.1 was drawn using these choices. Each data point on the graph represents a particular space-time event. The extreme upper right-hand point, for example, tells us that the puck was at a position 82 cm from the spot we called "zero" at 0.6 s after the time we called "zero."

This position-time graph is more than a picture; we can obtain information about the speed of the puck by looking at the steepness of the straight lines connecting the data points. Because the time intervals are all the same, the fastest average speed occurs when the puck covers the most distance in one of these time intervals. On the graph this happens when the vertical separation of two adjacent data points is the greatest. This causes the line connecting the two points to be the steepest. The steepest line occurs between 0.1 s and 0.2 s. The slowest average speed occurs during the first and last time intervals.

We measure the steepness of a line by calculating the ratio of the *rise* over the *run*, a quantity known as the *slope*.

$$\text{slope} = \frac{\text{rise}}{\text{run}}$$

The rise of the line is the difference between two vertical values, while the run is the difference between the two corresponding horizontal values. The slopes of the straight-line segments in Figure 2.1.1 represent the average speeds during each interval.

We can now use this concept to look at the average speed for the entire trip. In Figure 2.1.2 we have redrawn the graph and replaced the dot-to-dot lines with a single straight line. This is often done in graphing when additional information suggests it. In this case, it is reasonable to believe that the motion was not jerky, but smooth. We believe that the jerkiness in the data was due to uncertainties in measuring the positions of the puck and not in the motion of the puck itself. Although there are mathematical techniques for determining this line, we use the "eye-ball" method. Assuming that the measurements were too big as often as they were too small, the best straight line should have about half the data points above the line and half below the line as shown in Figure 2.1.2. In this graph the best straight line passes through the first and last data points, but this is not a requirement. In choosing the best line, it is possible to miss any given point, or even all the points.

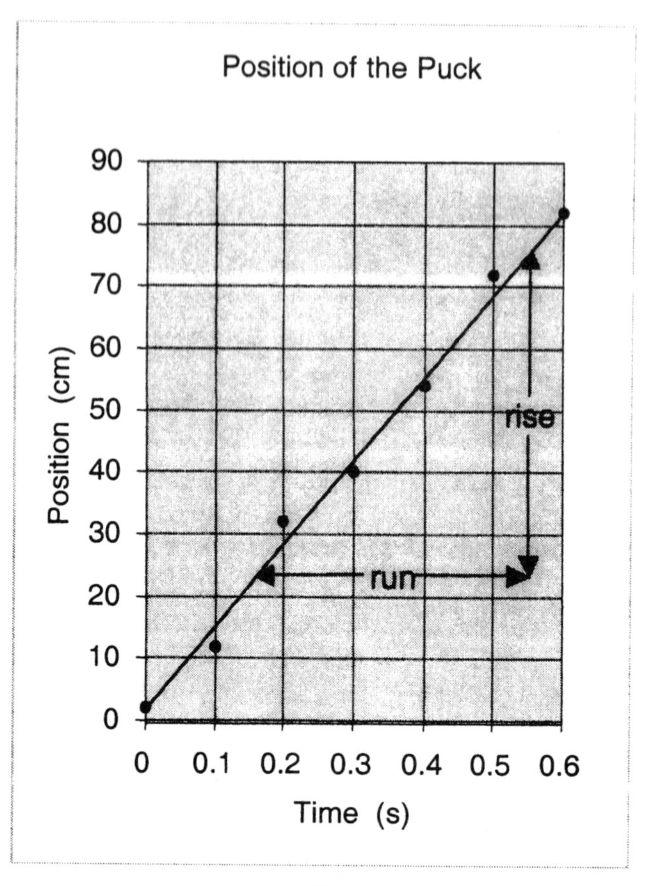

Fig. 2.1.2

We can use any two points on the straight line to calculate the slope of the line and obtain the average speed for the motion. It is more accurate to use widely separated points to reduce the effects of uncertainties in reading the values from the graph. Let's choose the points at 0.15 s and 0.55 s to calculate our slope. Then we have

$$\text{rise} = 75 \text{ cm} - 22 \text{ cm} = 53 \text{ cm}$$
$$\text{run} = 0.55 \text{ s} - 0.15 \text{ s} = 0.4 \text{ s}$$

$$\text{average speed} = \text{slope} = \frac{\text{rise}}{\text{run}} = \frac{53 \text{ cm}}{0.4 \text{ s}} = 133 \text{ cm/s}$$

(See Chapter 1 for our convention for rounding numbers in answers.)

Question: What would the graph look like if the puck were moving faster?
Answer: A puck with a faster speed would need to cover more distance in the same time. Therefore, the straight line would have a steeper slope.

2.2 Average Speed

The definition of average speed $\bar{s}$ is the distance traveled d divided by the time taken t.

$$\bar{s} = \frac{d}{t}$$

This expression can be used to obtain the average speed whenever we know the total distance traveled and the time required to do this.

Example 2.2.1

A family drove across the United States in a week. If they covered 4800 km, what was their average speed?

Our answer depends on the unit of time we use. Because we could use weeks, days, or hours, we could have several possible answers:

$$\bar{s} = \frac{4800\text{ km}}{1\text{ week}} = 4800\text{ km/week}$$

$$\bar{s} = \frac{4800\text{ km}}{7\text{ days}} = 686\text{ km/day}$$

$$\bar{s} = \frac{4800\text{ km}}{7 \times 24\text{ h}} = 28.6\text{ km/h}$$

Notice that these values include time for sleeping, eating, and rest stops.

Practice: If they drove 8 hours each day, what was the average speed during driving hours?
Answer: 85.7 km/h

Example 2.2.2

How far can you travel at an average speed of 85 km/h for 4.2 hours?

We can algebraically manipulate our definition for average speed to get an expression for the distance traveled by an object moving with this average speed for the given time. Begin by multiplying both sides of the definition by the time t and then canceling the t's on the right-hand side.

$$\bar{s}t = \frac{d}{\cancel{t}}\cancel{t}$$

Switching the two sides of the equation, we obtain our expression.

$$d = \bar{s}t$$

We now plug in the given values to get our answer.

$$d = \bar{s}t = \left(85\frac{\text{km}}{\text{h}}\right)(4.2\text{ h}) = 357\text{ km}$$

Practice: How far can an airplane fly in 3.5 hours at a speed of 180 km/h?
Answer: 630 km

Example 2.2.3

How long would it require to travel 730 km at an average speed of 80 km/h?

We can obtain the expression for the time required to travel a given distance at a given average speed by once again manipulating the definition for average speed. We begin by multiplying both sides by the time t as we did in the previous example. We then divide both sides by $\bar{s}$ and cancel the $\bar{s}$'s on the left-hand side.

$$\bar{s}t = d$$

$$\frac{\bar{s}t}{\bar{s}} = \frac{d}{\bar{s}}$$

We can now plug in the numbers to obtain our answer.

$$t = \frac{d}{\bar{s}} = \frac{730\text{ km}}{80\text{ km/s}} = 9.13\text{ h}$$

Note that dividing by the units km/h is equivalent to multiplying by h/km. We can then cancel the km's to get hours as the unit for our answer.

Practice: How long would it take a Cessna 172 to fly a distance of 500 km at an average speed of 175 km/h?
Answer: 2.86 h

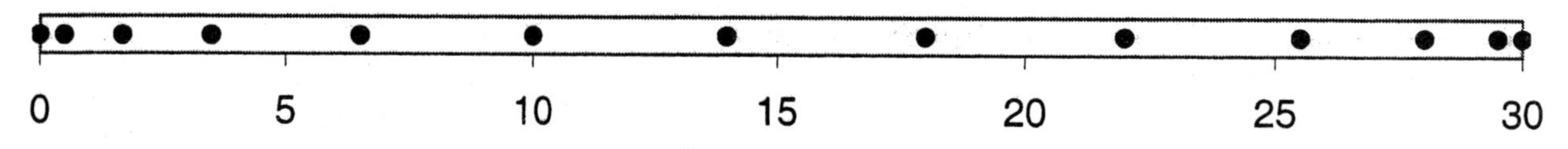

Fig. 2.3.1

2.3 Instantaneous Speed

Although the concept of an instantaneous speed is easy to comprehend, calculating it is another matter. The problem lies in the fact that dividing by an "instant of time" leads to mathematical confusion because dividing by zero is not defined in ordinary arithmetic. Isaac Newton invented the mathematics of calculus to deal with such problems. Fortunately, our graphical representation gives us an easy way of obtaining the instantaneous speed without resorting to calculus.

Consider the "strobe" drawing of a puck moving from left to right in Figure 2.3.1. The data from this drawing are given in Table 2.3.1. To determine the instantaneous speed of the puck at any instant during its motion, we start by making the position-time graph shown in Figure 2.3.2. Notice that the steepness of the line changes during the motion, indicating that the speed is varying; if it were constant, the curve would be a straight line. Portions of the curve are straight (or almost straight), indicating that parts of the trip had constant speeds. For instance, look at the end of the motion. The curve is straight and horizontal. The rise is zero, showing that the speed is zero. This can be verified by examining the last four entries in Table 2.3.1; the position of the puck doesn't change.

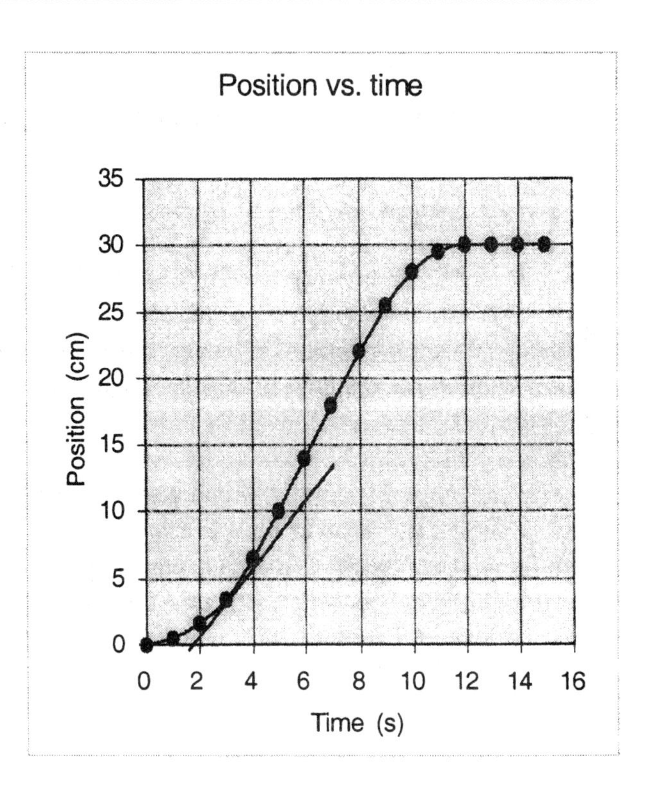

Fig. 2.3.2

The curve is almost horizontal at the beginning, indicating that the motion started out with a low speed. As the motion proceeds, the curve becomes steeper, indicating an increasing speed. The puck reaches its maximum speed between 5 s and 8 s, and then it slows down and stops.

Time (s)	Position (cm)	Av. Speed (m/s)
0	0.0	
		0.5
1	0.5	
		1.0
2	1.7	
		2.0
3	3.5	
		3.0
4	6.5	
		3.5
5	10.0	
		4.0
6	14.0	
		4.0
7	18.0	
		4.0
8	22.0	
		3.5
9	25.5	
		2.5
10	28.0	
		1.5
11	29.5	
		0.5
12	30.0	
		0.0
13	30.0	
		0.0
14	30.0	
		0.0
15	30.0	

Table 2.3.1

Question: Would a vertical line on this graph be possible? Why or why not?
Answer: Unlike the horizontal line, a vertical line does not represent a real motion because the object would have to be in many different positions at the same time, a physical impossibility. You could also think of it as representing an infinite speed, another impossibility.

How do we calculate the speed for a time when the curve is not straight? For example, consider again the motion of the puck represented by the graph in Figure 2.3.2. How fast was the puck moving at time t = 2 s? Graphically, we could find this speed by drawing a tangent to the curve at time $t = 2$ s and calculating the slope of this tangent line. When the curve is rather smooth, the tangent is a line that touches the curve at only one point. Why would the slope of the tangent line give us the instantaneous speed? To answer this question we must look "closely" at the graph in Figure 2.3.2. At time $t = 2$ s the graph is curved, indicating that the speed is changing. If we concentrate, however, on just the small portion of the graph near $t = 2$ s, say from $t = 1.99$ s to $t = 2.01$ s, and blow up this section of the graph (see Figure 2.3.3) the graph of position versus time appears straight, indicating uniform speed. Why is this? It takes time to change speed. Even the most expensive racecar requires some amount of time to speed up from zero to 60 mph. When we look at the motion of the puck for such a small interval of time, the puck has nearly uniform motion and the slope of the straight line is interpreted as the speed of the puck during that short interval. If we imagine extending this straight-line segment in both directions, it becomes the tangent to the curve at $t = 2$ s.

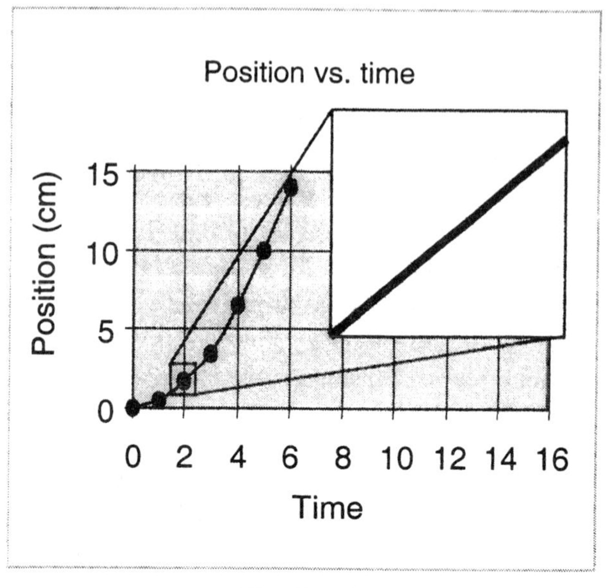

Fig. 2.3.3

Example 2.3.1

A tangent line has been drawn in Figure 2.3.2 for the time of 3 s. What is the slope of this line?

Let's choose the positions corresponding to $t = 2$ s and $t = 6$ s.

$$\text{slope} = \frac{\text{rise}}{\text{run}} = \frac{11\text{ cm} - 1\text{ cm}}{6\text{ s} - 2\text{ s}} = \frac{10\text{ cm}}{4\text{ s}} = 2.5\text{ cm/s}$$

Therefore the instantaneous speed at 3 s is about 2.5 cm/s.

Practice: Determine the instantaneous speed at t = 9 s.
Answer: You should have obtained something close to 3 cm/s.

2.4 Velocity-Time Graphs

Just as it was useful to draw the position-time graph to get a feeling for the speed of the object, it is useful to draw a velocity-time graph to develop a feeling for the acceleration. From the similarities of the definitions of average velocity and average acceleration, we know that the slope of the curve on the velocity-time graph gives us the magnitude of the acceleration. Because we have already calculated the average speeds for our motion in Table 2.3.1, let's use these values as approximations to the instantaneous speeds. Because we don't know when in the time intervals the average speeds actually occurred, a reasonable guess is to plot them at the middle of each interval in Figure 2.4.1.

During the first few seconds, not only was the speed increasing, but since the steepness of the velocity-time curve increases, the acceleration was also increasing. From 3 s to 7 s, the acceleration was getting smaller, but the speed was still increasing. After 7 s the acceleration was reversed and the speed decreased, eventually causing the puck to stop at about 13 s.

Question: Where did the acceleration have the maximum value in the forward direction? In the backward direction? Where was the acceleration equal to zero?

Answer: The steepest part of the rising curve occurs at about 2 s. Therefore, the maximum acceleration in the forward direction occurs at this time. The maximum value for the backward direction occurs between 10 and 12 s. For the acceleration to be zero, the slope of the tangent line must be zero. This occurs at the top of the curve near 7 s and near the end of the motion.

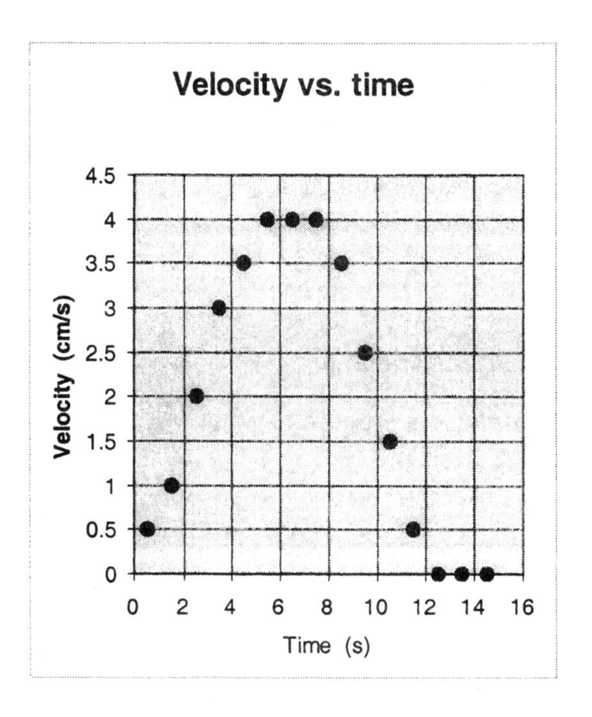

Fig. 2.4.1

2.5 Speed with Direction

Velocity is a vector quantity that includes direction as well as speed. In the case of motion along a straight line, we can describe the direction with a plus or a minus sign. Assume that we measure distances from west to east, as is done by the mile markers along Interstate highways. We then say that something moving east has a positive velocity and that something moving west has a negative velocity. We can then calculate the change in velocity by taking the difference of the final and initial velocities, including their signs.

Example 2.5.1

What is the change in velocity that occurs when a car traveling at 25 m/s east turns around and travels 15 m/s west?

Being careful to assign a plus sign to the eastward velocity and a minus sign to the westward velocity, we have

$$\Delta v = v_f - v_i = (-15\text{ m/s}) - (25\text{ m/s}) = -40\text{ m/s}$$

Because the answer is negative, the change in velocity points toward the west.

Practice: What is the change in velocity if the directions are reversed?
Answer: +40 m/s, where the plus sign means that the change in velocity points eastward.

2.6 Acceleration

In the text, we calculated the acceleration of a car that increased its speed from 40 mph to 60 mph in a time of 20 s.

$$\bar{a} = \frac{\Delta v}{\Delta t} = \frac{v_f - v_i}{t_f - t_i} = \frac{60\text{ mph} - 40\text{ mph}}{20\text{ s}} = \frac{20\text{ mph}}{20\text{ s}} = 1\text{ mph/s}$$

The car can also have a decrease in its speed. Suppose the car slowed down from 60 mph to 40 mph in the same 20 s. (Notice that this reverses the initial and final velocities.) We calculate the change in velocity in the same way by subtracting the initial velocity from the final velocity. Therefore, the car's average acceleration would be

$$\bar{a} = \frac{\Delta v}{\Delta t} = \frac{v_f - v_i}{t} = \frac{40\text{ mph} - 60\text{ mph}}{20\text{ s}} = \frac{-20\text{ mph}}{20\text{ s}} = -1\text{ mph/s}$$

The value of the acceleration is the same as it was in our first example except for the minus sign. At first glance it is tempting to say that the negative value shows up in the calculation because the object is slowing down. Although this is true in this case, the negative value really results from our choosing the direction of travel as the positive direction.

Velocity is a vector quantity that includes direction as well as speed, so we describe the direction with a plus or a minus sign. Assume that something traveling east has a positive velocity, while something traveling west has a negative velocity. We must then include the signs in the calculation to take care of these directions.

Example 2.6.1

What is the change in velocity of a car traveling westward that speeds up from 40 mph to 60 mph?

Adopting east as the positive direction, we have

$$\Delta v = v_f - v_i = -60\text{ mph} - (-40\text{ mph}) = -20\text{ mph}$$

This means that the car is traveling 20 mph faster in the negative direction, that is, westward. Note that this car would have an average acceleration of −1 mph/s if this change took place in 20 s.

Practice: What would the average acceleration be if the car were traveling eastward?
Answer: 1 mph/s

Mathematically, the direction of the acceleration arises from the vector nature of velocity. When we take the difference of two vectors, the answer is also a vector. Because acceleration is the difference of two velocity vectors divided by the time interval, acceleration is a vector quantity, and thus has a direction. As with velocity, the direction of the acceleration along a straight line is given by a plus or minus sign. The average acceleration in the last example is negative, indicating that the acceleration points westward.

But how do we know from the sign of the acceleration whether the car is speeding up or slowing down? In the text the car was moving in the positive direction and speeding up. When we calculated the acceleration, we obtained a positive value. Therefore, the acceleration was in the same direction as the velocity. In our last example, the car was traveling in the negative direction and speeding up, and we obtained a negative acceleration. When the velocity and acceleration vectors point in the same direction, the car speeds up. Conversely, when they point in opposite directions, the car slows down in agreement with our earlier calculation.

Example 2.6.2

Suppose that a ball is traveling at 10 m/s in a direction that we decide to call positive and that 6 s later it is moving in the opposite direction at 20 m/s. What is the average acceleration of the ball?

$$\bar{a} = \frac{\Delta v}{\Delta t} = \frac{v_f - v_i}{t} = \frac{-20 \text{ m/s} - 10 \text{ m/s}}{6 \text{ s}} = \frac{-30 \text{ m/s}}{6 \text{ s}} = -5 \text{ m/s}^2$$

Once again the minus sign tells us that the direction of the acceleration is in the negative direction. Because this is opposite to the initial velocity of the ball, the ball slows down. After the ball's speed is reduced to zero, the negative acceleration causes the ball to speed up in the negative direction.

It is standard to write the units of acceleration as m/s^2 rather than m/s/s, which is a bit ambiguous as to the order of division. We can see that these forms are the same if we remember that dividing by a fraction is the same as multiplying by the fraction turned over.

$$\text{m/s/s} = \frac{\text{m/s}}{\text{s}} = \frac{\text{m/s}}{\text{s/1}} = \frac{\text{m}}{\text{s}}\frac{1}{\text{s}} = \frac{\text{m}}{\text{s}^2}$$

Practice: What is the average acceleration of a car that changes its velocity of −20 m/s to a velocity of 20 m/s in 10 s?
Answer: 4 m/s^2

2.7 Free Fall: Making a Rule of Nature

Using our knowledge of how a ball behaves when dropped, we can fill in a table showing its acceleration and its speed at the end of each second. We know that the acceleration has a constant value throughout the fall. This causes the ball to speed up by 10 m/s each second. (The pattern is easier to see if we round off the acceleration.)

time (s)	**acceleration (m/s^2)**	**speed (m/s)**	**av. speed (m/s)**	**Δ position (m)**	**position (m)**
0	10	0			0
			5	5	
1	10	10			5
			15	15	
2	10	20			20
			25	25	
3	10	30			45

For any object, the distance traveled during a time interval is just the average speed during the time interval multiplied by the length of the time interval. For objects with a constant acceleration, the average speed is the average of the initial and final speeds.

$$\bar{s} = \frac{s_f + s_i}{2}$$

The average speeds and the changes in position are shown in the fourth and fifth columns of the table. Because we now know the change in position that occurs during each time interval, we can calculate the position of the object at the end of each second. This is shown in the last column.

Example 2.7.1

How far will the ball fall between $t = 3$ s and $t = 4$ s?

$$d = \bar{s}\,t = \frac{s_i + s_f}{2}\,t = \frac{(30\ \text{m/s}) + (40\ \text{m/s})}{2}(1\ \text{s}) = 35\ \text{m}$$

Practice: How far will the ball fall between $t = 3$ s and $t = 5$ s?
Answer: 80 m

2.8 Starting With an Initial Velocity

Until this point we have not had a relationship that allows us to calculate what happens if a falling object is initially in motion. We can use our definition of average acceleration to determine what happens to the velocity. By writing out Δv explicitly as the difference between the initial and final velocities, we have

$$a = \frac{\Delta v}{\Delta t} = \frac{v_f - v_i}{t}$$

where we've not used the average acceleration because the instantaneous acceleration is constant and equal to the average acceleration and we've used t for the time interval $t_f - t_i$. Rearranging terms yields

$$v_f = v_i + at$$

This form of the relationship tells us that the acceleration causes a fixed change in the velocity, and it doesn't matter if the initial velocity is zero or not.

Example 2.8.1

In the text we discussed a ball thrown into the air with an initial upward velocity of 20 m/s. What's the velocity of the ball after 2 s?

Choosing the upward direction to be positive, we have

$$v_f = v_i + at = 20\ \text{m/s} + \left(-10\ \text{m/s}^2\right)(2\ \text{s}) = 0$$

We used a negative value for the acceleration due to gravity because it points downward. Notice that this answer — that the ball has an instantaneous speed of zero — agrees with the verbal argument given in the text.

Practice: What is the velocity of the ball at 4 s?
Answer: –20 m/s. Note the symmetry between the upward motion and the downward motion.

The formula for the change in position when there's an initial velocity is a bit more complicated and we give it here without derivation.

$$d = v_i t + \tfrac{1}{2} a t^2$$

The first term on the right-hand side of the equation tells us how far the object would have gone without any acceleration, while the second term yields the distance for the special case of no initial velocity. The sum of the two distances gives us the total change in position.

Example 2.8.2

Returning to our example, what is the change in position for the thrown ball after 2 s?

$$d = v_i t + \tfrac{1}{2} a t^2$$
$$= (20\ \text{m/s})(2\ \text{s}) + \tfrac{1}{2}(-10\ \text{m/s}^2)(2\ \text{s})^2$$
$$= 40\ \text{m} - 20\ \text{m} = 20\ \text{m}$$

Again, each term in the equation represents different contributions to the ball's upward motion. The first term tells us the ball would have gone up 40 m without any gravity, but the effect of the acceleration due to gravity, expressed in the second term, reduced this by 20 m. The net result is that the ball rises 20 m in the first 2 s.

Example 2.8.3

What is the change in position after 4 s?

$$d = v_i t + \tfrac{1}{2} a t^2$$
$$= (20\ \text{m/s})(4\ \text{s}) + \tfrac{1}{2}(-10\ \text{m/s}^2)(4\ \text{s})^2$$
$$= 80\ \text{m} - 80\ \text{m} = 0$$

At first glance this answer might seem a bit surprising, because the ball has obviously traveled some distance. The key to understanding this result is realizing that d is the change in the position, or displacement, of the ball. Because the ball returned to the same position as it started, its change in position is zero. To get the total distance traveled, add the 20 m for the upward path to the 20 m for the downward path to get a total of 40 m.

Practice: Where is the ball at 3 s?
Answer: 15 m above the beginning height.

Problems

1. What is the average speed of a cruise ship that covers 370 km in 24 h?
2. What is the average speed of a jet transport that flies 8000 km in 10 h 12 min?
3. If a cheetah runs 1.5 miles in 2 min, what is its average speed in mph?
4. A biker travels 60 miles in 3 h. What is the biker's average speed in m/s?
5. In 1991 Carl Lewis broke the record for the 100-m dash in a time of 9.86 s. What was his average speed in mph?
6. In 1954 Roger Bannister was the first person to run a mile in less than 4 min. His time was 3 min 59.4 s. What was his average speed in mph?
7. Starting at 10 am, you bicycle at an average speed of 15 mph until noon when you enter the mountains. From noon until 4 pm you average only 10 mph. What is you average speed over the entire 6 h trip? (*Hint*: The average speed is not the average of the two speeds.)

8. You leave your home at noon hoping to get to your destination, which is 300 miles away, by 5 pm. During the first two hours of your trip you encounter road construction and only average 30 mph. What average speed do you need to maintain for the rest of your journey to get there on time?
9. If you can drive across Montana along Interstate 90 in 12 h at an average speed of 95 km/h, how far is it across Montana?
10. If Earth has an average orbital speed of 107,000 km/h, how far does it travel in one year?
11. It is 4470 km between New York City and Los Angeles. If a cyclist can ride for 8 h/day at an average speed of 25 km/h, how many days would it take to make the ride?
12. Assuming that a space ship traveling to Mars has to travel 300 million km at an average speed of 20,000 km/h, how many months would it take to make the journey?
13. Figure 2.P.1 is a position-time graph for the motion of a model rocket. What is the instantaneous speed of the rocket 8 s after launch?
14. Figure 2.P.1 is a position-time graph for the motion of a model rocket. What is the instantaneous speed of the rocket 3 s after launch?
15. A velocity-time graph for a ball falling in air is shown in Figure 2.P.2. What is the acceleration of the ball 8 s after it is dropped?
16. A velocity-time graph for a ball falling in air is shown in Figure 2.P.2. What is the acceleration of the ball 2 s after it is dropped?
17. A cheetah can obtain a speed of 72 km/h from a standing start in just 2 s. What is its average acceleration?
18. A horse accelerates from 4 mph to 20 mph in 4 seconds. What is the horse's average acceleration in m/s^2?

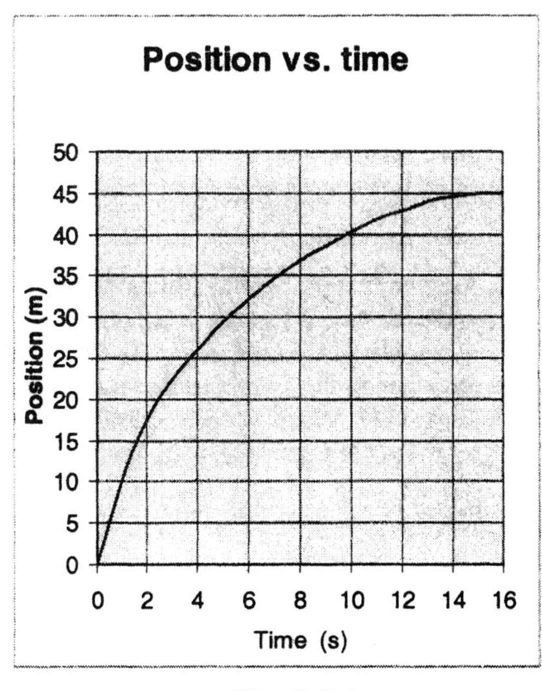

Fig. 2.P.1

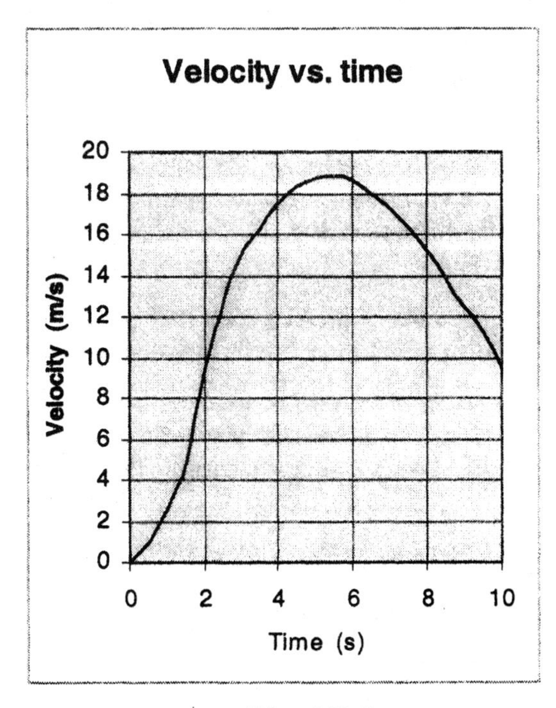

Fig. 2.P.2

19. In 1975 Don Garlits set the world's record for AA fuel dragsters by traveling 0.25 mile in 5.9 s from a standing start. He was traveling 254 mph at the end of the quarter mile.
(a) What was his average acceleration?
(b) What was his average speed?

20. A Ford Escort can cover 0.25 mile from a standing start in 19.4 s and be going 69 mph at the finish.
(a) What is its average acceleration?
(b) What is its average speed?

21. What is the average acceleration of a car that is initially traveling at 100 km/h north, but is traveling 80 km/h south 2 minutes later?

22. An airplane was flying north at 180 km/h. Ten minutes later it is flying south at the same speed. What was its average acceleration?

23. A ball is falling at 20 m/s just before it strikes the floor. It leaves the floor at a speed of 16 m/s traveling up. Find the ball's average acceleration if the collision lasts 0.04 s.

24. A baseball traveling at 80 mph is hit back at the pitcher with a speed of 110 mph. If the ball remains in contact with the bat for 0.01 s, what is the average acceleration of the ball?

25. What is the average acceleration of a ball that is thrown vertically upward on the Moon with a speed of 2 m/s if it takes 2.5 s to return?

26. A chunk of Martian rock falls off a cliff and drops 47.7 m in 5 s. What is the acceleration due to gravity on the surface of Mars?

27. Construct a table for the speed of a rock and the distance it has fallen at each second if is dropped from the top of a 123-m cliff.

28. Construct a table for the speed of a ball and its position above the ground at the end of each second if it is thrown upward at 39.2 m/s.

29. A ball is thrown vertically upward with an initial speed of 29.4 m/s.
(a) What is the value of the instantaneous speed when the ball reaches its maximum height?
(b) How long does it take to reach this height?
(c) How long does it take the ball to fall from the maximum height to the original height? (*Hint*: Remember that the upward motion is symmetric to the downward motion.)
(d) What is the maximum height reached by the ball?
(e) How fast is the ball going when it returns to its original height?

30. A stone is thrown straight upward at a speed of 19.6 m/s. Find the stone's speed and acceleration 2 s later.

31. A truck is traveling along a straight stretch of freeway at 25 m/s. How far will it travel in 10 s? If it accelerates at 2 m/s^2 to pass another truck, how far will it travel in the next 10 s?

32. A car is traveling along a straight stretch of freeway at 40 m/s. How far will it travel in 20 s? If it accelerates at 2 m/s^2 to slow down for a speed trap, how far will it travel in the next 5 s?

33. A ball traveling at 16 m/s falls past a window. How far will it fall during the next 2 s?

34. A ball is traveling upward at 30 m/s. How far will it travel during the next 2 s?

35. Assume that you are traveling 60 mph when you see a hazard ahead. If it takes you 1 s to recognize the hazard and apply the brakes, how far (in feet) will you travel during this second? If it takes you an additional 4 s to stop, how much farther will you travel?

3 — EXPLAINING MOTION

3.1 Adding Vectors

If two displacements are at right angles to each other, we can deter-mine the resultant displacement by applying the Pythagorean theorem. If we label the lengths of the three sides of a right triangle as shown in Figure 3.1.1, the Pythagorean theorem states that the square of the long side of the triangle is equal to the sum of the squares of the other two sides. Symbolically, we have

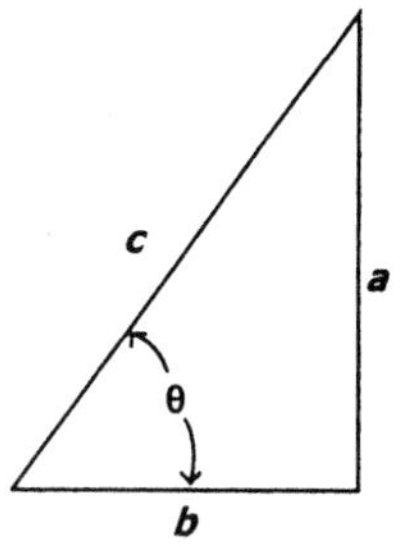

$$c^2 = a^2 + b^2$$

The direction of the resultant displacement can be measured with a protractor on a scale drawing. You can also obtain the angle if your calculator has trigonometric functions. The tangent of an angle is defined as the ratio of the lengths of the opposite and adjacent sides.

c
a
θ
b

Fig. 3.1.1

$$\tan\theta = \frac{\text{opposite}}{\text{adjacent}}$$

Therefore, divide the length of the side opposite the angle θ (in this case a) by the length of the adjacent side b. Then push the "inv" or "2nd" key followed by the "tan" key. The answer should appear in the calculator window.

Example 3.1.1

Assume that two forces at right angles to each other are applied to a ball. One force is 20 N to the north and the other is 30 N to the west. What is the resultant force on the ball?

We use the Pythagorean theorem to obtain the square of the net force.

$$c^2 = a^2 + b^2 = (20\text{ N})^2 + (30\text{ N})^2 = 1300\text{ N}^2$$

We now take the square root to obtain the force.

$$c = \sqrt{1300\text{ N}^2} = 36.1\text{ N}$$

To obtain the angle, we divide the westhward force by the northward force to get 0.15 and punch the appropriate keys on the calculator to get θ = 56.3°west of north.

Practice: What is the net force if the northward force is increased to 40 N?
Answer: 50 N at an angle of 36.9° west of north

3.2 The Second Law

Newton's second law can be used to find the net force required to accelerate a given mass with a given acceleration. The direction of the net force is the same as the direction of the acceleration. The relationship can also be rearranged algebraically to solve for the acceleration or the mass.

Example 3.2.1

What net force is required to accelerate a 200-kg box with an acceleration of 4 m/s² south?

$$F_{net} = ma = (200\text{ kg})(4\text{ m/s}^2) = 800\text{ kg}\cdot\text{m/s}^2 = 800\text{ N}$$

Therefore, the force must have a size of 800 N and act toward the south, the same direction as the acceleration.

Practice: What net force must be acting on a cyclist with a mass of 60 kg, if the cyclist has an acceleration of 2 m/s² to the north?
Answer: 120 N north

Example 3.2.2

What acceleration would result from a net force of 200 N acting westward on a person with a mass of 80 kg?

We begin with Newton's second law and divide both sides by the mass m to obtain

$$\frac{F_{net}}{m} = \frac{\cancel{m}a}{\cancel{m}}$$

Therefore,

$$a = \frac{F_{net}}{m} = \frac{200\text{ N}}{80\text{ kg}} = 2.5\frac{\text{N}}{\text{kg}} = 2.5\frac{\text{km}\cdot\text{m/s}^2}{\text{kg}} = 2.5\text{ m/s}^2$$

toward the west.

Practice: What would the acceleration be if the force were increased to 300 N?
Answer: 3.75 m/s² westward

Example 3.2.3

A crate with an unknown mass undergoes an acceleration of 3.5 m/s² when a net force of 1900 N is applied to it. What is the mass of the crate?

We begin with Newton's second law and divide both sides by the acceleration a to obtain

$$\frac{F_{net}}{a} = \frac{m\cancel{a}}{\cancel{a}}$$

or

$$m = \frac{F_{net}}{a} = \frac{1900\ \text{N}}{3.5\ \text{m/s}^2} = 543\frac{\text{kg}\cdot\text{m/s}^2}{\text{m/s}^2} = 543\ \text{kg}$$

Practice: What would the mass be if it required 2300 N to produce the same acceleration?
Answer: 657 kg

Example 3.2.4

A Volkswagen with a small driver has a mass of 1000 kg and can accelerate from rest to 22.4 m/s (50 mph) in 7.9 s. What is the net force on the VW?

The second law gives us the net force if we know the mass and the acceleration. Because the mass is given, we begin by finding the VW's average acceleration.

$$\bar{a} = \frac{\Delta v}{\Delta t} = \frac{v_f - v_i}{\Delta t} = \frac{(22.4 - 0)\ \text{m/s}}{7.9\ \text{s}} = 2.84\ \text{m/s}^2$$

We can now calculate the average net force exerted on the car.

$$F_{net} = ma = (1000\ \text{kg})(2.84\ \text{m/s}^2) = 2840\ \text{N}$$

If our driver gets out and four football players get into the car, the total mass climbs to 1420 kg. We would then expect the average acceleration to decrease. If we assume that the VW's driving force doesn't change, we can calculate the new acceleration.

$$a = \frac{F_{net}}{m} = \frac{2840\ \text{N}}{1420\ \text{kg}} = 2\ \text{m/s}^2$$

The power of mathematics allows us to determine much more about this situation. For example, even though we haven't actually found four willing football players, we can calculate various other values, such as the time it would take the VW to reach the same final speed. We rearrange the definition of acceleration to determine this time.

$$\Delta t = \frac{\Delta v}{a} = \frac{(22.4 - 0)\ \text{m/s}}{2\ \text{m/s}^2} = 11.2\ \text{s}$$

These calculations can be tested in a real experiment. The point here is that our rules allowed us to make statements about the VW's performance before the experiment is done.

Practice: What would the average acceleration be if two 100-kg football players got out?

Answer: This reduces the mass to 1220 kg, and the average acceleration becomes 2.33 m/s^2.

3.3 Weight

Calculating the weight of an object under the influence of Earth's gravitational attraction is an application of Newton's second law. The weight is the force that produces a downward acceleration equal to the acceleration due to gravity **g**, which has a constant value of 9.8 m/s^2 near Earth's surface.

Example 3.3.1

What is the weight of a football tackle with a mass of 150 kg?

$$W = mg = (150\text{ kg})(9.8\text{ m/s}^2) = 1470\text{ N} \qquad (330\text{ lb})$$

Practice: What is the weight of a ballerina with a mass of 46 kg?
Answer: 451 N

Example 3.3.2

What is the mass of a car that has a weight of 15,000 N?

Solving our equation for the mass, we have

$$m = \frac{W}{g} = \frac{15{,}000\text{ N}}{9.8\text{ m/s}^2} = 1530\frac{\text{N}}{\text{m/s}^2} = 1530\frac{\text{kg}\cdot\text{m/s}^2}{\text{m/s}^2} = 1530\text{ kg}$$

Practice: What is the mass of a person with a weight of 850 N?
Answer: 86.7 kg

Example 3.3.3

What is the weight of an 80-kg person standing on Mars where the acceleration due to gravity is 3.7 m/s^2?

$$W = mg = (80\text{ kg})(3.7\text{ m/s}^2) = 296\text{ N}$$

Compare this with a weight of 784 N on Earth.

3.4 Drawing Free-Body Diagrams

The first step in solving many mechanics problems is to correctly identify all the forces acting on the object in question. We do this with a free-body diagram. The first force that we typically draw on our free-body diagram is the gravitational force $W_{\text{Earth, object}}$. We then ask ourselves if there are any other "non-contact" forces, forces that can act on our object without physically touching it (like magnetic forces). After we've accounted for all non-contact forces, all other forces must be due to agents that are touching the object. These contact forces are either pushes or pulls. A pull is usually exerted on an object by a string or a rope, and we call these tension forces. A push can either be a normal force that acts perpendicular to the surface between the pusher and the pushee, or a frictional force that acts parallel to this surface. If you push down on a table at some arbitrary angle, you will feel two sensations; pressure on your palm, and the skin being pulled back on your bones. The pressure on your palm is due to the *normal* (or perpendicular) part of this push and the pulling on your skin is due to the frictional force. If you are unsure of whether a push is normal or frictional, imagine greasing the surface and ask, "Can I still apply the force?" If you answer yes, the push is a normal force.

Example 3.4.1

A paper business card of mass 25 g is held to a refrigerator door (as shown below) by a magnet with a mass of 50 g. Sketch a free-body diagram for the *business card* and a separate free-body diagram for the *magnet.* In each case:

- draw vectors to indicate the forces that are exerted on the object,
- label and describe each force,
- identify the object exerting each force and the object on which that force is exerted.

The force vectors on your diagrams should be consistent with any known relative magnitudes.

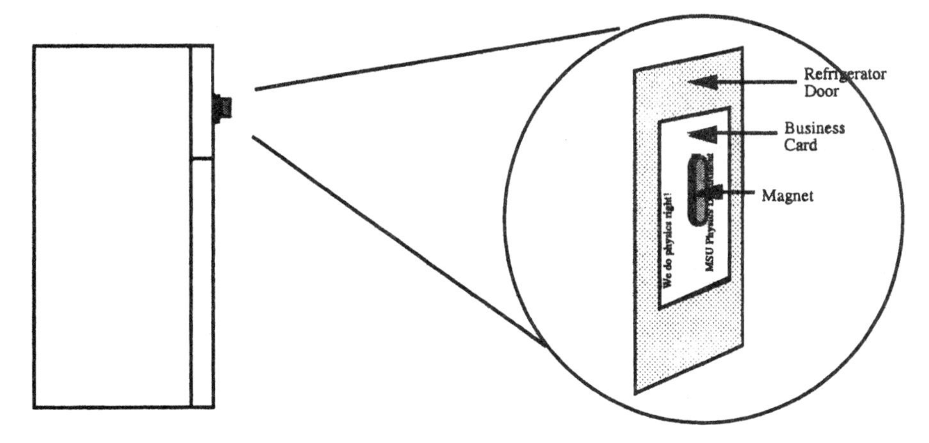

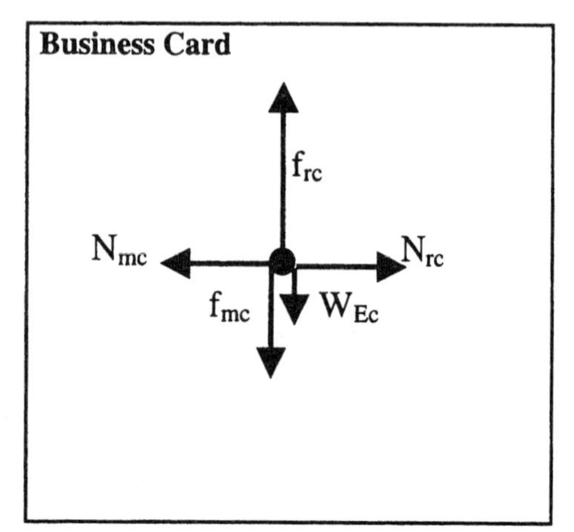

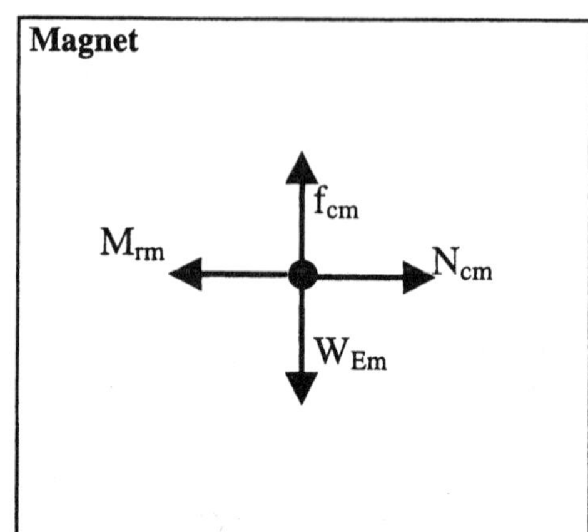

E – earth
c – card
m – magnet
r – refrigerator

3.5 Friction

To a good approximation, the kinetic frictional force acting on an object is proportional to the force keeping the object in contact with the surface. This force is called the *normal* force because it acts perpendicular to the surface. This statement can be written symbolically as $F = \mu N$, where μ is the *coefficient of kinetic friction.* The value of μ depends on the types of surfaces, but is insensitive to the normal force and the speed of the object. For a horizontal surface with no other vertical forces acting, the normal force is the weight of the object.

Example 3.5.1

What is the frictional force acting on a 50-kg crate in motion if the coefficient of kinetic friction is 0.1?

$$F = \mu N = \mu mg = (0.1)(50\text{ kg})(9.8\text{ m/s}^2) = 49\text{ N}$$

Example 3.5.2

If a force of 99 N is applied to the crate in the previous example, what is the acceleration of the crate?

$$a = \frac{F_{net}}{m} = \frac{(99\text{ N} - 49\text{ N})}{50\text{ kg}} = 1\text{ m/s}^2$$

Example 3.5.3

If we observe the crate in the previous examples accelerating at 2 m/s², what must be the value of the applied force?

The net force is given by Newton's second law.

$$F_{net} = ma = (50\text{ kg})(2\text{ m/s}^2) = 100\text{ N}$$

We must push with a force that overcomes the sliding friction and still yields a net force of 100 N. Therefore, the applied force is 149 N.

Practice: What acceleration would a 199-N force produce?
Answer: The net force of 199 N – 49 N = 150 N would give an acceleration of 3 m/s².

3.6 Using Free-Body Diagrams

In science fiction movies it is often convenient to assume the existence of a "universal translating device" to allow communication between two worlds. In our study of mechanics, Newton's second law is the translator between the forces on our free-body diagram (dynamics) and the motion that results (kinematics). Sometimes we know the magnitude of all the forces on our diagram and we can use Newton's second law to find the acceleration. At other times we know something about the acceleration and we can use Newton's second law to find the magnitudes of some of the forces.

Example 3.6.1

A horizontal string pulls a 25-kg block across a level table. The tension in the string is 125 N. If the coefficient of kinetic friction between the block and the table is $\mu_k = 0.3$, find the acceleration of the block.

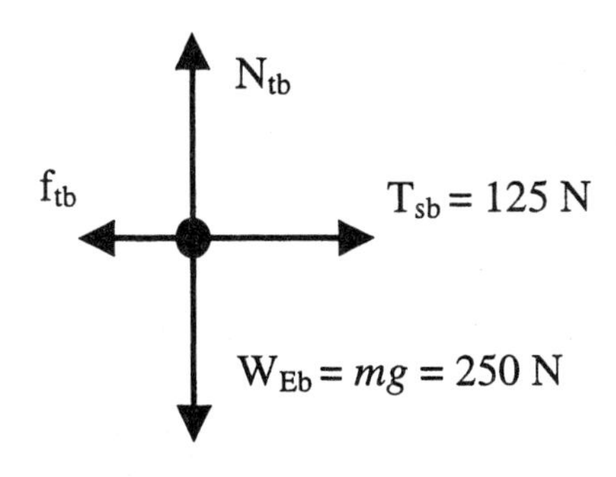

We begin by drawing a free-body diagram for the block. We know some of the forces acting on the block, but not all of them. We know, however, that the block is not accelerating up or down, so the force up (the normal force) must balance the force down. The normal force is also 250 N. We have used our knowledge of the acceleration to find one of the forces. The friction force can now be calculated; it is just the coefficient of kinetic friction times the normal force: $f = \mu_k N = (0.3)(250\text{ N}) = 75\text{ N}$. Because we now know all the forces acting on the block, we can find the acceleration of the block using Newton's second law. The block has 125 N acting to the right and only 75 N acting to the left. The net (or unbalanced) force 50 N to the right. This net force accelerates the block:

$$a = \frac{F_{net}}{m} = \frac{50\text{ N}}{25\text{ kg}} = 2\text{ m/s}^2$$

3.7 Newton's Third Law

A can sits on a table. The free-body diagram for the can is simple. The gravitational force acts downward and the table pushes upward with a normal force, as shown at the right. Newton taught us that every force has a companion force that is equal in magnitude and opposite in direction. What is the third-law force for the normal force exerted on the can by the table? We are tempted to point to the gravitational force acting on the can. Indeed the normal force acting upward is equal and opposite to the gravitational force acting downward, but these forces do not comprise a third-law pair. Third-law forces are always part of an interaction between two objects. If the action force is a normal force, the reaction force must also be a normal force. They also never act on the same object. If A pushes on B, then B pushes back on A just as hard. If I hit your chin with my fist, your chin hits my fist with a force equal in strength and opposite in direction. One of the forces hurts your chin. The other force hurts my fist. The companion force to N_{tc} must also be a normal force. If the table pushes up on the can, the can must also push down on the table. The companion force N_{ct} does not appear on our diagram for the can. It is a force on the table and therefore appears on a free-body diagram for the table.

N_{tc}

W_{Ec}

Example 3.7.1

Find the third-law companion force for the gravitational force W_{Ec}.

It is a gravitational force exerted by the can on Earth W_{cE}.

If the can and table are inside an elevator that is accelerating upward, the normal force exerted upward on the can must be greater in magnitude than the gravitational force exerted downward. The two forces on our free-body diagram are no longer equal in magnitude. Third-law forces, however, remain equal and opposite. The table cannot push up on the can with a greater force than the can pushes down on the table.

Problems

1. Two horizontal forces act on a toy boat; 50 N to the north and 50 N to the west. What is the net horizontal force (magnitude and direction) on the boat?

2. What is the net force (magnitude and direction) produced by the following forces; 4 N acting toward the east, 9 N acting toward the south, and 6 N acting toward the north?

3. A force of 40 N acts to the right and a force of 60 N acts downward. What force is needed to produce no acceleration?
4. A light suspended from the ceiling has a weight of 120 N and is being pulled sideways with a force of 20 N. At what angle from the vertical does the suspending cord hang?
5. What net force is needed to accelerate a car at 3 m/s^2 east if it has a mass of 2400 kg?
6. An airplane has an acceleration of 2 m/s^2 on its take-off roll. If the airplane has a mass of 1100 kg, what is the net force acting on the airplane?
7. What acceleration is produced by an 882-N force toward the right acting on a 90-kg person?
8. A 90-kg person has a gravitational force of 150 N on the Moon. What would the free-fall acceleration on the Moon be?
9. A model train undergoes an acceleration of 0.4 m/s^2 when a net force of 3.2 N is applied to it. What is the mass of the train?
10. What is the mass of an ice skater, if a net force of 95 N results in an acceleration of 1.7 m/s^2?

*11. The same force is applied to two colored objects. The blue one accelerates at 10 m/s^2 and the red one at 30 m/s^2. What is the ratio of their masses? Which one has the smaller mass?

*12. The same force is applied to two toy cars. The milk truck accelerates at 3 m/s^2, while the sports car accelerates at 9 m/s^2. If the sports car has a mass of 0.2 kg, what is the mass of the milk truck?

*13. A book sliding along a horizontal table has a constant frictional force of 4 N acting on it. If it has a mass of 1 kg and an initial speed of 8 m/s, how long will it take to come to rest?

*14. An ice skater with a mass of 50 kg moves with a constant speed of 8 m/s in a straight line. How long will it take a force of 100 N to stop the skater if it were applied so as to oppose the motion? What would the speed of the skater be if the force were applied for twice as long?

15. A 300-N crate drops straight down from a helicopter into a large snowdrift. The crate slows as it crashes through the snow. (a) Is the upward force exerted by the snow on the crate greater than, equal to, or less than 300 N? (b) What is the direction of the crate's acceleration? (c) Are your answers to (a) and (b) consistent?
16. A 1500-kg car accelerates uniformly from rest to 30 m/s in a time of 20 s. What is the net force acting on the car?
17. A seismograph is taken to Mars to see if there are any "marsquakes." If it has a mass of 12 kg on Earth, what are its mass and weight on Mars. The acceleration due to gravity on Mars is 3.7 m/s^2.
18. A child has a mass of 32 kg and a weight of 314 N when measured on Earth. What weight and mass would the child have on the Moon where the gravitational force is one-sixth that on Earth?
19. A man with a mass of 85 kg is riding in an elevator that is accelerating upward at 1.2 m/s^2. How much force does the scales exert on him?
20. A woman with a mass of 70 kg is riding in an elevator that is accelerating downward at 0.8 m/s^2. How much force does the scales exert on her?
21. How much lighter does a 600-N woman feel in an elevator accelerating downward at 1 m/s^2?
22. How much heavier does a 900-N man feel in an elevator accelerating upward at 1 m/s^2?
23. Draw a free-body diagram for a 700-N woman standing on a floor.
24. Draw a free-body diagram for an 800-N parachutist at a time when the air resistance is 200 N.

25. Draw a free-body diagram for the light described in Problem 4.

26. A 20-kg block is accelerating to the right. The block experiences an applied force of 300 N to the right and a frictional force of 100 N to the left. Draw a free-body diagram for the block.

*27. You are pulling a crate at constant speed across the floor of a service elevator while the elevator is accelerating upward at 1 m/s^2. If the crate's mass is 100 kg and the coefficient of sliding friction is 0.4, how hard are you pulling?

*28. You are pulling a crate at constant speed across the floor of a service elevator while the elevator is accelerating downward at 1.5 m/s^2. If the crate's mass is 100 kg and the coefficient of sliding friction is 0.4, how hard are you pulling?

29. You are applying a force of 650 N in an attempt to budge a 110-kg desk, which is resting on the floor. The coefficient of static friction between the desk and the floor is 0.80. What is the magnitude of the frictional force?

30. A person in free fall with a mass of 80 kg experiences a wind resistance of 500 N. What is the acceleration of the person?

31. A parachutist has a mass of 70 kg and an acceleration of 7 m/s^2. How large is the air resistance?

32. What is the frictional force acting on a 5.4-kg wagon if it accelerates at 1.5 m/s^2 under an applied force of 12 N?

33. A rope is used to pull a 20-kg block across the floor with an acceleration of 3 m/s^2. If the tension in the rope is 160 N, what is the coefficient of sliding friction between the floor and the block?

34. A team of 10 sled dogs is pulling a 200-kg sled across a level field of snow. The coefficient of sliding friction between the sled and the snow is 0.1 and the acceleration of the sled is 1.5 m/s^2. With what force is each dog pulling, assuming that they share the burden equally and all pull straight ahead?

4 — MOTIONS IN SPACE

4.1 Acceleration Revisited

The average acceleration of an object is the change in velocity—the final velocity minus the initial velocity—divided by the time taken to make the change. We can use the graphical techniques developed in the text to obtain the size and direction of the change in velocity and average acceleration. Let's begin with motion along a straight line.

Example 4.1.1

A car was observed traveling east at 10 m/s. During the next minute the car increased its speed to 30 m/s. What were the car's change in velocity and average acceleration?

Let's agree to take east as the positive direction. Thus the change in velocity is

$$\Delta v = v_f - v_i = 30 \text{ m/s} - 10 \text{ m/s} = +20 \text{ m/s}$$

We've inserted the plus sign to emphasize that the change in velocity points directly east. It is always a good idea to obtain the result using a vector diagram. (We show the change in velocity as a dashed arrow.)

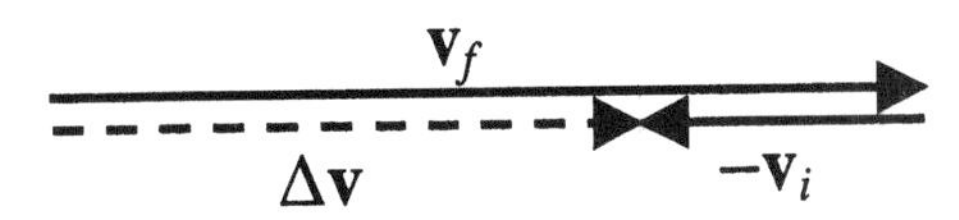

We can now calculate the average acceleration of the car.

$$\bar{a} = \frac{\Delta v}{\Delta t} = \frac{+20 \text{ m/s}}{60 \text{ s}} = +0.333 \text{ m/s}^2$$

Once again the plus sign indicates that the average acceleration points directly east.

Practice: What were the change in velocity and average acceleration if the car slowed down from 30 m/s to 10 m/s?
Answer: –20 m/s and –0.333 m/s^2

Example 4.1.2

A car was observed traveling east at 30 m/s. After one minute the car was observed traveling west at 10 m/s. What were the car's change in velocity and average acceleration?

Let's agree to take east as the positive direction. Thus the change in velocity is

$$\Delta v = v_f - v_i = -10 \text{ m/s} - 30 \text{ m/s} = -40 \text{ m/s}$$

The minus sign indicates that the change in velocity points directly west. It is always a good idea to obtain the result using a vector diagram. (We show the change in velocity as a dashed arrow.)

$-\mathbf{v}_i$ $\mathbf{v}_f$

$\Delta\mathbf{v}$

We can now calculate the average acceleration of the car.

$$\bar{a} = \frac{\Delta v}{\Delta t} = \frac{-40 \text{ m/s}}{60 \text{ s}} = -0.667 \text{ m/s}^2$$

Once again the minus sign indicates that the average acceleration points directly west.

Practice: What are the change in velocity and average acceleration if the car was initially traveling west at 10 m/s and then traveling east at 30 m/s?
Answer: +40 m/s (directly east) and +0.667 m/s^2

If the two velocities do not point in the same or opposite directions, we obtain the change in velocity using a scale drawing. The magnitude is measured using a ruler and the direction using a protractor. If the two velocities are at right angles to each other, we can apply the Pythagorean theorem developed in Section 3.1, *Adding Vectors*, to mathematically determine the acceleration.

Example 4.1.3

Assume that a car is moving north at 35 mph and, some time later, it is moving west at the same speed. What is the average acceleration of the car if the change takes place in 60 s?

We begin by finding the difference in the two velocities. Because they are at right angles to each other, we use the Pythagorean theorem.

$$c^2 = a^2 + b^2 = (35 \text{ mph})^2 + (35 \text{ mph})^2 = 2450 \text{ mph}^2$$

$$c = \sqrt{2450 \text{ mph}^2} = 49.5 \text{ mph}$$

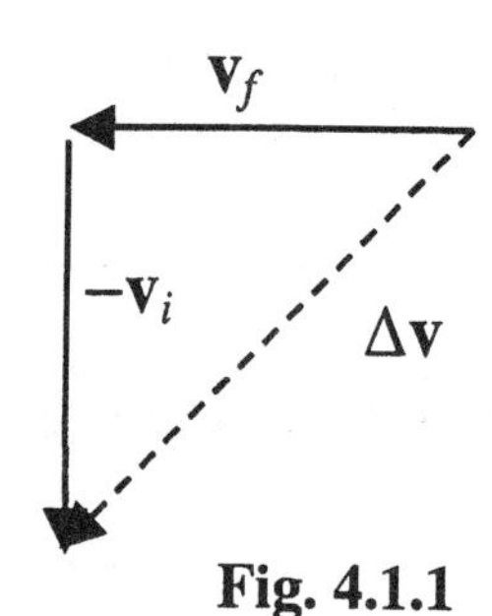

Fig. 4.1.1

A scale drawing of these velocities like the one in Figure 4.1.1 shows that the angle of the change in velocity is directly southwest. Or, you can use a calculator (see Section 3.1) to find that the angle is 45°. However, you will still need a sketch to determine the general direction.

The value of the acceleration is obtained by dividing this change in velocity by the 60 s it took for the change to occur. Therefore, remembering the vector nature of acceleration, our answer is

$$\bar{\mathbf{a}} = \frac{\Delta \mathbf{v}}{\Delta t} = \frac{49.5 \text{ mph SW}}{60 \text{ s}} = 0.825 \text{ mph/s SW}$$

Practice: What is the average acceleration if the final speed is 20 mph?
Answer: 0.672 mph/s at 60.3° south of west

4.2 Acceleration in Circular Motion

We can determine the size of the centripetal acceleration by examining the force needed to move the ball in a circle. The simple apparatus illustrated in Figure 4.2.1 gives a way of measuring the force. Select a ball and a weight and tie them to opposite ends of a string that you have threaded through a tube. As you twirl the ball, adjust its motion to keep the hanging weight at a constant height. The hanging weight supplies the centripetal force that is necessary to cause the ball to travel in a circle at a constant speed.

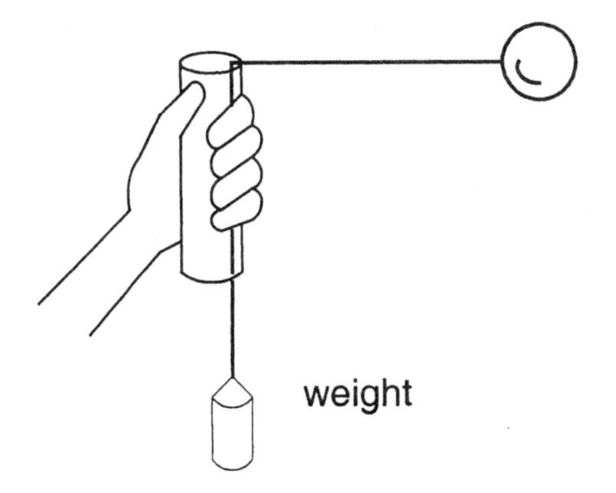

Fig. 4.2.1

If you now time how long it takes the ball to complete one revolution and measure the radius of the circle to determine its circumference, you can calculate the ball's speed. By doubling the value of the hanging weight you can investigate the dependence of the centripetal force on radius and speed. Return to the same radius as before and obtain the new speed. Then use the original speed and measure the new radius. You can complete your investigation by trying other values of the hanging weight and varying the mass of the ball. After you do all this, you would find that the magnitude of the centripetal force is given by the mass times the speed squared divided by the radius of the circular path.

$$F = \frac{mv^2}{r}$$ *centripetal force*

Because the centripetal force is the only horizontal force acting on the ball, we can use this expression for the force in Newton's second law.

$$F = \frac{mv^2}{r} = ma$$

Canceling the mass, we obtain an expression for the centripetal acceleration.

$$a = \frac{v^2}{r}$$ *centripetal acceleration*

Example 4.2.1

A cyclist turns a corner with a radius of 20 m at a speed of 6 m/s. If the mass of the cyclist and cycle is 80 kg, what centripetal force is required?

We begin by calculating the centripetal acceleration.

$$a = \frac{v^2}{r} = \frac{(6\text{ m/s})^2}{20\text{ m}} = 1.8\text{ m/s}^2$$

We can now insert this value into Newton's second law to get the centripetal force.

$$F_{net} = ma = (80\text{ kg})(1.8\text{ m/s}^2) = 144\text{ N}$$

Practice: What centripetal force is required if the speed is reduced to 3 m/s?
Answer: 36 N

Example 4.2.2

Suppose you are able to rotate a bucket of water in a vertical circle at a constant speed. The acceleration of the water at the top of the circle is perpendicular to its velocity, or straight down. The water, however, does not fall on your head if the bucket is moving quickly around the circle. Because the acceleration is perpendicular to the velocity, it causes the water to change direction without changing speed.

Let's rotate this bucket fairly quickly, such that it makes one full rotation in only 1.2 s. If the mass of the water is 2 kg and the radius of the circle is 0.8 meter, find the normal force exerted by the bucket on the water at the top of the circle.

First, let's find the acceleration of the water. If the water travels a distance

$$2\pi r = 2\pi(0.8\text{ m}) = 5.03\text{ m}$$

in 1.2 s, its speed is 4.19 m/s. The centripetal acceleration is therefore

$$a = \frac{v^2}{r} = \frac{(4.19\text{ m/s})^2}{0.8\text{ m}} = 21.9\text{ m/s}^2$$

down toward the center of the circle. (Note that this is the acceleration of the water; we do not need to add in the gravitational acceleration.) The net force acting on the water is

$$F_{net} = ma = (2\text{ kg})(21.9\text{ m/s}^2) = 43.8\text{ N}$$

Water

$W_{Earth,water}$

$N_{bucket,water}$

A free-body diagram for the water at the top of the circle includes only two forces; the gravitational force of Earth on the water and the normal force of the bucket on the water. (Although the sides of the bucket are pushing on the water, these forces cancel and leave us with a net force downward on the water.)

The gravitational force is just mg = 19.6 N, so the normal force must equal 43.8 N − 19.6 N = 24.2 N.

As the bucket is rotated slower and slower, the net force required to keep the water in its circular path decreases and so the normal force decreases. How slow could we swing the bucket before the water spills out and lands on our head?

The water leaves the bucket when the normal force between the bucket and the water goes to zero. At this point, the net force is just the gravitational force of 19.6 N. The speed would therefore be

$$v = \sqrt{\frac{F_{net}r}{m}} = \sqrt{\frac{(19.6\text{ N})(1\text{ m})}{(2\text{ kg})}} = 3.13\text{ m/s}$$

or about one rotation every 1.6 s.

4.3 Projectile Motion

The key idea in solving problems involving projectile motion is to remember that the vertical and horizontal motions are independent of each other as long as we can neglect the air resistance. The horizontal motion is very simple because there is no acceleration in the horizontal direction. Therefore, the horizontal component of the velocity v_h is constant, and the horizontal distance R traveled is given by

$$R = v_h t$$

where t is the time the ball is in flight.

The vertical motion is just free fall. Therefore, we can use the two equations for motion with an initial velocity developed in Section 2.8.

$$v_{vf} = v_{vi} + a_v t$$

$$d = v_{vi}t + \tfrac{1}{2}a_v t^2$$

where d is the change in vertical position, v_{vi} and v_{vf} are the initial and final vertical components of the velocity, respectively, and a_v is the acceleration due to gravity with its sign chosen to agree with the choice of direction in the problem.

Notice that these equations are connected because they both contain the time t. Depending on the problem, we will need to work with one direction to obtain the time, and then use it for the other direction. Typically, we solve for the time using the vertical motion.

Example 4.3.1

A ball is thrown from a 78.4-m high cliff with a horizontal velocity of 5 m/s. How long is the ball in the air, and where does it land on the flat plain below?

To calculate the time t, we look at the vertical motion. We do this by using our formula for the position of the ball.

$$d = v_{vi}t + \tfrac{1}{2}a_v t^2$$

Let's assume that we call the downward direction positive. Setting $v_{vi} = 0$, $a_v = g$, and $d = H$, we have

$$H = \tfrac{1}{2}gt^2$$

Solving this equation for t^2, we can substitute in the given values.

$$t^2 = \frac{2H}{g} = \frac{2 \times 78.4\ \text{m}}{9.8\ \text{m/s}^2} = 16\ \text{s}^2$$

Therefore, $t = 4$ s.

We can now use this time to calculate the distance that the ball travels horizontally.

$$R = v_h t = (5\ \text{m/s})(4\ \text{s}) = 20\ \text{m}$$

Practice: How do your answers change if the horizontal speed is increased to 6 m/s?

Answer: Because the vertical and horizontal motions are independent, the time the ball is in the air remains the same. The increased horizontal speed means that the ball will travel farther during this time, $R = 24$ m.

Example 4.3.2

Professor Kirkpatrick hits his golf ball such that it leaves the ground moving 50 m/s at an angle of 37° with respect to the level ground. If we ignore the effects of air resistance, how far will the ball travel downrange?

First, we should note that the ball would travel farthest if it was hit at an angle of 45°, but Professor Kirkpatrick chose 37° to make the problem easier. At 37°, the velocity and its components form a 3-4-5 right triangle. If the magnitude of the velocity is 50 m/s, then the vertical component is 30 m/s and the horizontal component is 40 m/s.

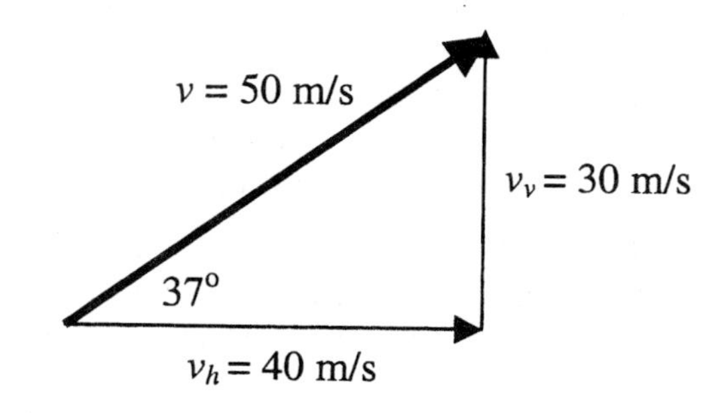

The most important thing to remember about projectile motion is that the vertical motion is independent of the horizontal motion. The horizontal motion is simple. An object in free-fall has no acceleration in the horizontal direction, so the horizontal part of its velocity never changes. From the moment the ball leaves the club to the moment it hits the ground, it will move 40 m downrange for every second that elapses.

$$R = v_h t$$

We just need to know how long the ball remains in the air. This is determined entirely by the vertical component. The ball remains in the air for the same time as a ball thrown straight up with an initial velocity of 30 m/s. We know that after one second, such a ball is moving at a speed of roughly 20 m/s (the ball slows down by 10 m/s each second). It takes three seconds for it to reach the top and another three seconds to return to the ground. The golf ball is in the air for a total of six seconds. The range would therefore be 240 meters! (Note: In practice the interaction between the air and the golf ball is very important and may not be ignored.)

The comparison between Narang and Earth discussed in the text can be seen mathematically in the expression for the vertical position of the dart. In this case, the acceleration due to gravity is downward, and we've put the minus sign in explicitly to make the outcome more obvious.

$$y = v_{vi}t - \frac{1}{2}gt^2$$

The first term on the right-hand side gives the height of the dart if there were no acceleration like the conditions on Narang. The vertical component of the velocity v_{vi} is constant, and the dart climbs by the same amount each second. If this were the only term, the dart would travel along a straight line. The second term is just the distance any object falls under the influence of gravity if it starts from rest. The combination of the two terms gives the behavior on Earth. Therefore, the dart shot at the gorilla falls away from the straight line exactly as the gorilla falls away from the branch, and the dart will hit its target. Notice that this argument does not depend on the value of v_i. This means that the speed of the dart doesn't matter as long as the dart hits the gorilla before the gorilla hits the ground.

Does this arrangement depend on the value of g? That is, will you get the same result if you perform this experiment on the Moon were $g_m = 1.6 \text{ m/s}^2$?

Problems

1. Find the size and direction of the change in velocity for each of the following initial and final velocities.
   (a) 3 m/s west to 6 m/s west.
   (b) 6 m/s west to 3 ms/s west.
   (c) 3 m/s west to 6 m/s east.
2. What is the change in velocity for each of the following initial and final velocities?
   (a) 50 km/h forward to 100 km/h forward.
   (b) 50 km/h forward to 100 km/h backward.
3. An airplane was initially flying north with a speed of 100 mph. Later it was observed flying west at the same speed. What was its change in velocity?
4. A car is driving north at 120 km/h when it turns so that it is driving east at 120 km/h. What is the change in velocity of the car?
5. Find the size and direction of the change in velocity for an initial velocity of 3 m/s south and a final velocity of 6 m/s west.

6. A bus was initially traveling with a velocity of 50 km/s west. Later it was observed traveling with a velocity of 100 km/h south. What was the change in velocity of the bus?

7. An object experiences an average acceleration of 4 m/s^2 west for a time of 2 s. Find the object's final velocity (magnitude and direction) given the following initial velocities. (a) 3 m/s west. (b) 6 m/s east.

8. An object experiences an average acceleration of 3 m/s^2 east for a time of 2 s. Find the object's initial velocity (magnitude and direction) given the following final velocities. (a) 8 m/s east. (b) 2 m/s east.

9. Io, one of Jupiter's many moons, orbits the giant planet in a nearly circular orbit of radius 422,000 km. The period for the orbit is 1 day, 18 hours, and 28 minutes. What is Io's average centripetal acceleration?

10. The Hubble Space Telescope orbits Earth at an altitude of approximately 550 km above Earth's surface. The satellite orbits Earth about once every 95 minutes. (a) What is the radius of the satellite's orbit? (b) What is the speed of the satellite in m/s? (c) What is the centripetal acceleration of the satellite?

11. The top of a loop of a roller coaster has a radius of 10 m. If the cars are traveling at 12 m/s, what force does the seat exert on a 80-kg passenger?

12. A 90-kg stunt pilot pulls out of a dive while flying at 50 m/s. If he flies a circular path with a radius of 200 m, what force does his seat exert on him at the bottom of the path?

13. A football is thrown with a horizontal speed of 15 m/s and a vertical speed of 12 m/s upward. What are these speeds 1 s later?

14. What are the horizontal and vertical speeds of the football in the previous question 2 s after it is thrown?

15. A ball is thrown horizontally at a speed of 20 m/s over level ground. (a) If it starts out at a height of 19.6 m, how long does it take for the ball to reach the ground? (b) How far will it move horizontally in this time?

16. A bowling ball rolls off a giant's table at a speed of 20 m/s and lands 60 m from the base of the table. (a) For how many seconds is the ball falling? (b) How high is the table?

17. A ball is launched horizontally at a speed of 40 m/s. With what speed will it hit the ground if it falls for 3 s?

18. A stunt car drives off the top of a parking garage at 30 m/s. How fast will it be going just before it hits that crash pad if it falls for 2 s?

19. A football is thrown with a vertical speed of 9.8 m/s and a horizontal speed of 15 m/s. How far down field should the wide receiver be to catch this pass?

20. A human cannonball is launched so that he has horizontal and vertical speeds of 9.8 m/s. (a) How long does it take him to reach the top of his path? (b) How far away should he place the safety net?

5 — GRAVITY

5.1 Newton's Gravity

We can calculate the acceleration of the Moon using the expression for centripetal acceleration developed in Section 4.2. In addition to the radius r of the Moon's orbit, we need to know the Moon's speed. The average speed v can be computed from the distance traveled during one revolution divided by the time it takes to complete one revolution; that is, the Moon's period. If we make the very good approximation that the Moon's orbit is circular, we can obtain the circumference C of the orbit.

$$C = 2\pi r = 2\,(3.14)\,(3.84 \times 10^8 \text{ m}) = 2.41 \times 10^9 \text{ m}$$

Next we need the time for the Moon's period. In order to end up with units of m/s^2, we convert the time to seconds.

$$T = (27.3 \text{ days})\left[\frac{24 \text{ hrs}}{1 \text{ day}}\right]\left[\frac{60 \text{ min}}{1 \text{ hr}}\right]\left[\frac{60 \text{ s}}{1 \text{ min}}\right] = 2.36 \times 10^6 \text{ s}$$

Therefore,

$$v = \frac{d}{t} = \frac{C}{T} = \frac{2.41 \times 10^9 \text{ m}}{2.36 \times 10^6 \text{ s}} = 1.02 \times 10^3 \text{ m/s}$$

And, finally, we can calculate the centripetal acceleration a.

$$a = \frac{v^2}{r} = \frac{\left(1.02 \times 10^3 \text{ m/s}\right)^2}{3.84 \times 10^8 \text{ m}} = 2.71 \times 10^{-3} \text{ m/s}^2$$

5.2 The Law of Universal Gravitation

Example 5.2.1

What is the gravitational force exerted on Earth by the Sun?

Using the masses of Earth M_e and the Sun M_s and the Earth-Sun distance R_{es} given in the Appendix, we have

$$F = G\frac{M_e M_s}{R_{es}^2}$$

$$F = \left(6.67 \times 10^{-11}\,\frac{\text{N}\cdot\text{m}^2}{\text{kg}^2}\right)\frac{\left(5.98 \times 10^{24} \text{ kg}\right)\left(1.99 \times 10^{30} \text{ kg}\right)}{\left(1.50 \times 10^{11} \text{ m}\right)^2} = 3.53 \times 10^{22} \text{ N}$$

Although this is a very large force, the mass of Earth is so large that the force produces a very small acceleration.

$$a = \frac{F}{m} = \frac{3.53 \times 10^{22}\ \text{N}}{5.98 \times 10^{24}\ \text{kg}} = 5.90 \times 10^{-3}\ \text{m/s}^2$$

Practice: What acceleration does the Sun have due to Earth's gravitational force?
Answer: $1.77 \times 10^{-8}\ \text{m/s}^2$

5.3 The Value of G

We can obtain the mass of Earth by equating the weight of an object of mass m near Earth's surface with the gravitational force obtained from Newton's law of universal gravitation.

$$F = mg = \frac{GmM_e}{R_e^2}$$

where M_e is Earth's mass and R_e is Earth's radius. Canceling the object's mass m from both sides of the equation and solving for Earth's mass, we obtain

$$M_e = \frac{gR_e^2}{G} = \frac{\left(9.8\ \text{m/s}^2\right)\left(6.37 \times 10^6\ \text{m}\right)^2}{6.67 \times 10^{-11}\ \text{N}\cdot\text{m}^2/\text{kg}^2} = 5.96 \times 10^{24}\ \text{kg}$$

The slight difference between this value and the one in the Appendix is due to round off.

5.4 Gravity Near Earth's Surface

We can calculate the value of g near Earth's surface using the law of universal gravitation and the values for Earth's mass and radius.

$$g = \frac{F}{m} = G\frac{M_e m}{R_e^2 m} = G\frac{M_e}{R_e^2}$$

$$= \left(6.67 \times 10^{-11}\ \frac{\text{N}\cdot\text{m}^2}{\text{kg}^2}\right)\left(\frac{5.98 \times 10^{24}\ \text{kg}}{\left(6.37 \times 10^6\ \text{m}\right)^2}\right) = 9.83\frac{\text{N}}{\text{kg}} = 9.83\ \text{m/s}^2$$

5.5 Satellites

Because the orbits of many Earth-orbiting satellites are nearly circular and we know how to calculate the centripetal forces and accelerations for circular paths (Section 4.2), we'll restrict ourselves to circular orbits. This allows us to calculate the force exerted on a satellite of mass m in two ways. We begin by writing the expression for the centripetal force required to maintain a satellite in a circular orbit of radius r with a speed v.

$$F = \frac{mv^2}{r}$$

Then we write the expression for the gravitational force exerted by Earth on the satellite.

$$F = G\frac{mM_e}{r^2}$$

We can set these two expressions for the force equal to each other because the gravitational force is the centripetal force needed to make the satellite stay in its orbit.

$$\frac{mv^2}{r} = G\frac{mM_e}{r^2}$$

After canceling an *m* and one *r* on each side of the equation, we arrive at an expression for the square of the speed of the satellite in a circular orbit of radius *r*.

$$v^2 = G\frac{M_e}{r}$$

Note that this result does not depend on the mass of the satellite. All satellites with this orbital radius must have this speed.

Now that we have the speed of the satellite, we can calculate how long it takes to go around once, a time known as its *period T*. Because the satellite must travel a distance equal to the circumference *C* of the circular orbit, its period is given by

$$T = \frac{C}{v} = \frac{2\pi r}{v}$$

To obtain an alternate expression for the period, we can replace *v* by our previous result to obtain

$$T = 2\pi\sqrt{\frac{r^3}{GM_e}}$$

Notice that this equation does not contain the mass of the satellite. Therefore, all satellites orbiting the same central body obey the relationship

$$T^2 = kr^3$$

where *k* is a constant. This was first discovered by Kepler and is now known as Kepler's third law. This expression tells us that the period of the satellite increases as the orbital radius increases. This makes sense as we already know that the Moon has a lot longer period than a space shuttle.

Example 5.5.1

What speed must a satellite have in order to be in orbit just above the surface of Earth?

We can use our expression if we neglect the obvious air resistance.

$$v^2 = G\frac{M_e}{r} = \left(6.67 \times 10^{-11}\,\frac{\text{N}\cdot\text{m}^2}{\text{kg}^2}\right)\left(\frac{5.98 \times 10^{24}\ \text{kg}}{6.37 \times 10^6\ \text{m}}\right) = 6.26 \times 10^7\ \text{m}^2/\text{s}^2$$

We obtain v by taking the square root.

$$v = 7910\ \text{m/s} = 7.91\ \text{km/s}$$

Note that this is consistent with the value of 8 km/s used in the text.

Practice: What is the speed of a satellite in a circular orbit 500 km above the surface?
Answer: 7620 m/s

Example 5.5.2

How long would it take a satellite near Earth's surface to orbit Earth?

$$T = \frac{2\pi r}{v} = \frac{2(3.14)(6.37 \times 10^6\ \text{m})}{7.91 \times 10^3\ \text{m/s}^2} = 5060\ \text{s} = 84.3\ \text{min}$$

Practice: What is the period (in minutes) of a satellite with an altitude of 500 km?
Answer: 94.4 min

Example 5.5.3

What is the altitude of a geosynchronous satellite that has a period of 24 hours?

We begin by taking our equation for the period

$$T = 2\pi\sqrt{\frac{r^3}{GM_e}}$$

and solving it for the radius. Divide both sides by 2π and then square the result.

$$\frac{T^2}{4\pi^2} = \frac{r^3}{GM_e}$$

Multiplying both sides by GM_e, we obtain

$$r^3 = \frac{GM_eT^2}{4\pi^2}$$

$$= \frac{\left(6.67 \times 10^{-11}\,\frac{\text{N}\cdot\text{m}^2}{\text{kg}^2}\right)(5.98 \times 10^{24}\ \text{kg})(8.64 \times 10^4\ \text{s})^2}{39.5} = 7.54 \times 10^{22}\ \text{m}^3$$

Taking the cube root using a calculator yields

$$r = 4.22 \times 10^7 \text{ m} = 42{,}200 \text{ km}$$

This is the distance above the center of Earth, so we must subtract the radius of Earth to obtain the altitude of 35,800 km.

5.6 The Field Concept

The gravitational field at a point in space is defined as the force an object would experience at that point, divided by the mass of the object. It is the force per unit mass. Because the field has the units of acceleration, the vector symbol **g** is often used to represent it. Once we know the field, the force (including its direction) acting on an object of mass *m* located at that point is just *m***g**.

Example 5.6.1

What is the gravitational field $\mathbf{g}_s$ of the Sun at the location of Earth?

The size of the field can be calculated using Earth as the test object and dividing the gravitational force on Earth by Earth's mass. Because this is exactly the calculation that we did near the end of Example 5.2.1, the result is 5.9×10^{-3} N/kg directed toward the Sun. It is important to realize that we could have used any object for this calculation. The value of the field would be the same.

We can also calculate this value directly.

$$g_s = \frac{F_e}{M_e} = G\frac{M_s M_e}{R_{es}^2 M_e} = G\frac{M_s}{R_{es}^2}$$

$$= \left(6.67 \times 10^{-11} \frac{\text{N} \cdot \text{m}^2}{\text{kg}^2}\right)\frac{1.99 \times 10^{30} \text{ kg}}{\left(1.5 \times 10^8 \text{ m}\right)^2} = 5.9 \times 10^{-3} \frac{\text{N}}{\text{kg}}$$

Practice: What is the gravitational field due to Earth at the location of the Moon?
Answer: 2.72×10^{-3} N/kg as calculated in Section 5.1

Problems

Astronomical data are provided in the Appendix.

1. What is the acceleration of Earth in its orbit about the Sun? How does this compare to the Moon's acceleration about the Sun?
2. What is the centripetal acceleration of Mars about the Sun?

3. What is the centripetal acceleration of Neptune about the Sun?
4. Saturn's largest moon Titan orbits Saturn once every 15 hours 22 minutes at a mean radius of 1,220,000 km. What is Titan's centripetal acceleration about Saturn?
5. What gravitational forces does Venus exert on the Sun?
6. What gravitational force does the Sun exert on Jupiter?
7. How does the maximum gravitational force that Jupiter exerts on Earth compare to the gravitational force the Sun exerts on Earth?
8. How does the gravitational force Earth exerts on the Moon compare to the average gravitational force the Sun exerts on the Moon?
9. Two objects released from rest in a region of space where there is no net external gravitational force accelerate toward each other. If one object accelerates 9 times as much as the other, how do their masses compare?
10. Two objects in outer space accelerate toward each other under the influence of their mutual gravitational attraction. If one object has four times the mass of the other, how do their accelerations compare?
11. What is the gravitational force that the Moon exerts on 1 kg of sea water on the side of Earth nearest the Moon? What is the force on a different 1 kg of sea water on the opposite side of Earth from the Moon?
12. What is the gravitational force on a 1-kg mass in the space shuttle when it is 300 km above Earth's surface?

*13. The mass of Mars is 0.1 times that of Earth and the radius of its orbit is 1.5 times that of Earth. How does Mars's gravitational force on the Sun compare to that of Earth on the Sun?

*14. The radius of Venus's orbit is 0.7 times that of Earth and its mass is 0.8 times Earth's mass. How does the Sun's gravitational force on Venus compare to the Sun's force on Earth?

*15. Jupiter is 5.2 times as far away from the Sun as Earth and it has 318 times the mass of Earth. Will Jupiter exert more or less force on the Sun than Earth does?

*16. Given that the average distance of the Moon from the Sun is 390 times the distance between Earth and the Moon and that the mass of the Sun is 33,300 times that of Earth, what is the ratio of the forces exerted on the Moon by Earth and by the Sun?

17. The acceleration due to gravity on the surface of Mars is 3.73 m/s^2. Given that Mars's radius is 3400 km, calculate Mars's mass.
18. A 1-kg block on the Moon has a weight of 1.62 N. Given that the Moon's radius is 1740 km, calculate the Moon's mass.
19. An astronaut on a strange planet has a mass of 65 kg and a weight of 240 N. What is the value of the acceleration due to gravity on this planet?
20. In a science fiction story several astronauts visit a strange planet known as K3 that is orbiting a nearby star. To determine the gravitational acceleration on K3 they drop a 2-kg block from a height of 10 m and measure the fall time to be 4 s. What is the value of the gravitational acceleration on K3? What is the weight of the block?
21. The solar system's largest moon, Ganymede, has a radius of 2600 km and a mass of 1.5×10^{23} kg. What is the acceleration due to gravity on the surface of the Ganymede?
22. Mars has two tiny moons. The smaller one, Deimos, has an average radius of about 7 km and mass of 1.9×10^{15} kg. What is the acceleration due to gravity on the surface of Deimos? The maximum height that you can jump is inversely proportional to the gravitational acceleration. If you can jump about 1 m on Earth, how high could you jump on Deimos?

*23. The mass of Venus is 80% that of Earth while its radius is 95% that of Earth. What is the value of the acceleration due to gravity on Venus?

*24. The radius of Mars is about 0.5 times Earth's radius, and its mass is 0.1 times Earth's mass. What would you expect for the value of the acceleration of gravity on the surface of Mars?

25. What is the orbital velocity of a satellite with an orbital radius equal to twice Earth's radius?

26. What is the orbital velocity of a satellite orbiting the Moon near its surface?

27. What is the orbital period of the satellite with an orbital radius equal to twice the radius of Earth?

28. What is the orbital period of the satellite orbiting the Moon near its surface?

*29. Mercury has a rotational period of 58.6 days. How high would a satellite have to be in order to orbit Mercury once each Mercurian day?

*30. If the rotational period of Mars is 24.6 h, how high would a satellite have to be in order to orbit Mars once each Martian day?

31. Use the fact that the Moon orbits Earth once every 27.3 days at an average distance that is about 60 times the radius of Earth to calculate Earth's mass.

32. At the geosynchronous distance, a satellite orbits Earth once every 24 h. How many times per day does a satellite orbit Earth if it is at one quarter the geosynchronous distance?

33. What is the value of Earth's gravitational field at a distance equal to twice Earth's radius?

34. What is the value of the Sun's gravitational field at its surface?

6 — MOMENTUM

6.1 Changing an Object's Momentum

An object's momentum is changed by an impulse, a force acting for a time interval, according to the expression that we obtained from Newton's second law.

$$\Delta(m\mathrm{v}) = \mathrm{F}_{net}\,\Delta t$$

When using this expression, we must remember that momentum and impulse are both vectors. For the case of motion along a straight line, we can once again use plus and minus signs to indicate direction.

Example 6.1.1

What is the change in the momentum of a freely falling ball with a mass of 1 kg during each second of fall?

We obtain the change in momentum by looking at the impulse exerted on the ball by the force of gravity; that is, the ball's weight.

$$\Delta(mv) = F_{net}\,\Delta t = (9.8\ \mathrm{N})(1\ \mathrm{s}) = 9.8\ \mathrm{N\cdot s} = 9.8\ \mathrm{kg\cdot m/s}$$

Because we know that the velocity of an object in free fall changes by 9.8 m/s during each second, we can check our result by calculating the change in momentum directly. Because the mass of the ball doesn't change, the change in momentum is simply given by the mass times the change in velocity.

$$\Delta(mv) = m\Delta v = (1\ \mathrm{kg})(9.8\ \mathrm{m/s}) = 9.8\ \mathrm{kg\cdot m/s}$$

Example 6.1.2

A 0.2-kg ball falls to the pavement and rebounds. If the ball's speed is 20 m/s just before it hits the ground and 15 m/s just after it leaves the ground, what are the ball's change in momentum and impulse?

Let's choose up to be the positive direction. We can then calculate the change in momentum, being careful to take the directions of the velocities into account.

$$\Delta p = p_f - p_i = m(v_f - v_i) = (0.2\ \mathrm{kg})(15\ \mathrm{m/s} - (-20\ \mathrm{m/s})) = +35\ \mathrm{kg\cdot m/s}$$

The change in momentum is directed upward. Because the impulse is equal to the change in momentum, the impulse is also 35 kg·m/s upward.

Example 6.1.3

A 1500-kg car is initially traveling at 30 m/s. If the frictional force is constant at 10,000 N, how long will it take the car to stop?

We can solve our equation for the time Δt and plug in the given values to obtain our answer.

$$\Delta t = \frac{\Delta(mv)}{F_{net}} = \frac{m(v_f - v_i)}{F_{net}} = \frac{(1500\text{ kg})(0 - 30\text{ m/s})}{-10{,}000\text{ N}} = 4.5\text{ s}$$

Note that we used a negative value for the frictional force as it was acting in the direction opposite to the velocity.

Practice: How long would it take the car to stop if the maximum frictional force were only 5,000 N?
Answer: 9 s

6.2 Conservation of Linear Momentum

Conservation of linear momentum holds whenever there is no net external force acting on a system. When this condition is met, the total momentum of all parts of the system must be the same before and after an interaction. In computing the total momentum of a system, we must be very careful to include the direction of each momentum. For motion along a straight line, the direction can be indicated by a plus or minus sign.

Example 6.2.1

Let's assume that a 50-kg woman is standing still on a 10-kg giant skateboard. If she then walks to the right at 1 m/s, what is the resulting velocity of the skateboard?

Begin by choosing the direction to the right as positive. Then set the total final momentum of the woman and the skateboard equal to the initial momentum, which is zero in this case.

$$m_b v_b + m_w v_w = 0$$

where the subscripts b and w refer to the board and woman, respectively. We can now solve the equation for the velocity of the board and plug in the given values.

$$v_b = -v_w \frac{m_w}{m_b} = (-1\text{ m/s})\frac{50\text{ kg}}{10\text{ kg}} = -5\text{ m/s}$$

where the minus sign indicates that the board is moving in the negative direction; that is, to the left. This answer makes sense; because the board has one-fifth the mass, it needs 5 times the speed to have the same size momentum as the woman.

Practice: What is the velocity of the board if the woman is replaced by a man with a mass of 80 kg?
Answer: –8 m/s

Example 6.2.2

Let's repeat the previous example for the case when the board and the woman have an initial velocity of 6 m/s to the right.

We once again set the final momentum equal to the initial momentum. Let's use the subscript i for the initial value. The woman's velocity is now 1 m/s larger than that of the board, or 7 m/s.

$$m_b v_b + m_w v_w = (m_b + m_w) v_i$$

$$v_b = \frac{(m_b + m_w) v_i - m_w v_w}{m_b} = \frac{(60 \text{ kg})(6 \text{ m/s}) - (50 \text{ kg})(7 \text{ m/s})}{10 \text{ kg}}$$

$$= \frac{10 \text{ kg} \cdot \text{m/s}}{10 \text{ kg}} = 1 \text{ m/s}$$

Therefore, the board slows down to 1 m/s, but it is still traveling to the right.

Practice: What is the final velocity of the board if the initial velocity is only 4 m/s?
Answer: –1 m/s

Example 6.2.3

Let's look at a collision of two air carts on a frictionless, horizontal air track. Assume that the carts have equal masses (m = 0.2 kg) and that one of the carts is initially stationary. The moving cart has a velocity to the right (v = 0.5 m/s). What is the total initial momentum of the two carts?

The total initial momentum of the two carts is just the vector sum of the individual initial momenta. Let's call momenta to the right positive.

$$p_i = m_1 v_1 + m_2 v_2 = mv + 0 = (0.2 \text{ kg})(0.5 \text{ m/s}) = 0.1 \text{ kg} \cdot \text{m/s}$$

Is momentum conserved if the first cart stops and the second cart leaves with velocity v?

The total final momentum in this case is

$$p_f = m_1 v_1 + m_2 v_2 = 0 + mv = (0.2 \text{ kg})(0.5 \text{ m/s}) = 0.1 \text{ kg} \cdot \text{m/s}$$

Momentum is conserved in this collision.

Is momentum conserved if both carts leave the collision with one-half the original velocity?

$$p_f = m_1 v_1 + m_2 v_2 = m\tfrac{1}{2}v + m\tfrac{1}{2}v = mv$$
$$= (0.2\text{ kg})(0.25\text{ m/s}) + (0.2\text{ kg})(0.25\text{ m/s}) = 0.1\text{ kg}\cdot\text{m/s}$$

Therefore, momentum is also conserved for this possibility.

Is momentum conserved if the cart that was initially moving rebounds in the opposite direction with $\frac{1}{2}v$ and the struck cart leaves with $\frac{3}{2}v$?

$$p_f = m_1 v_1 + m_2 v_2 = -m\tfrac{1}{2}v + m\tfrac{3}{2}v = mv$$
$$= (0.2\text{ kg})(-0.25\text{ m/s}) + (0.2\text{ kg})(0.75\text{ m/s}) = 0.1\text{ kg}\cdot\text{m/s}$$

Therefore, momentum is also conserved for this third possibility. Note that the velocity of the first ball is negative in the calculation to reflect that its motion is in the opposite direction. In the next chapter we will learn which of these possibilities actual happen in collisions of air carts.

6.3 Investigating Accidents

Example 6.3.1

A 1500-kg car traveling at 30 m/s undergoes a head-on collision with a 4500-kg truck traveling in the opposite direction at 20 m/s. If they wind up as one big heap, what was the velocity of the two vehicles immediately after the collision?

Let's begin by calculating the total momentum p_i before the crash. Choosing the direction of the truck's velocity as positive, we have

$$p_i = m_c v_c + m_t v_t$$
$$= (1500\text{ kg})(-30\text{ m/s}) + (4500\text{ kg})(20\text{ m/s}) = 45{,}000\text{ kg}\cdot\text{m/s}$$

This must be equal to the final momentum p_f of the total mass.

$$P_f = (m_c + m_t)\,v_f = p_i$$

Solving for v_f we obtain

$$v_f = \frac{p_i}{m_c + m_t} = \frac{45{,}000\text{ kg}\cdot\text{m/s}}{6000\text{ kg}} = 7.5\text{ m/s}$$

The two vehicles move in the truck's original direction because the truck had the larger initial momentum.

6.4 Two-Dimensional Collisions

Example 6.4.1

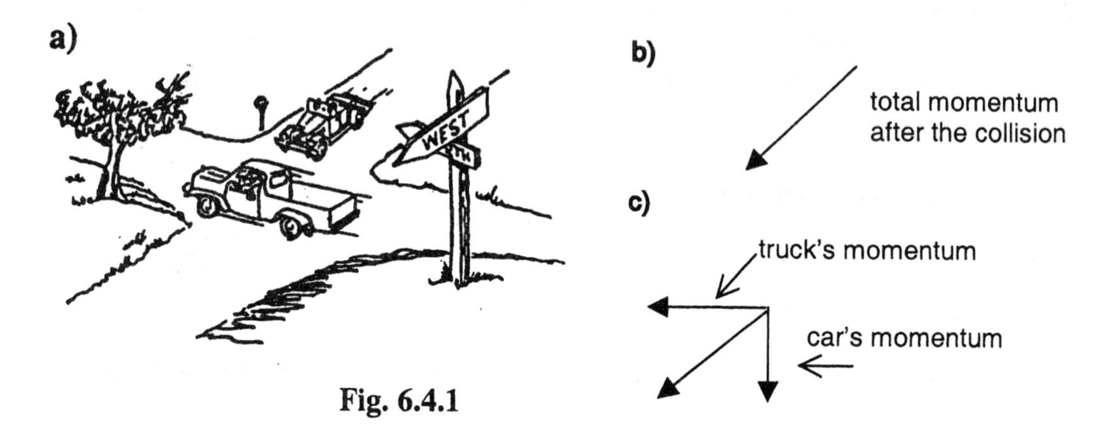

Fig. 6.4.1

Suppose a westbound 1000-kg car collides with a northbound 2000-kg pickup truck, as shown in Figure 6.4.1(a). The direction of the total momentum just after the collision is shown in the diagram of Figure 6.4.1(b). If the total momentum after the collision was 50,000 kg·m/s, how fast was each vehicle moving before the collision?

We make a scale drawing where 0.5 cm represents 10,000 kg·m/s. We know that the momenta of the two cars before the crash must add together as vectors to give their total momentum after the crash. This can only be done in one way. The initial momenta must be the west and north components of the final momentum, as shown in Figure 6.4.1(c). Measurement of the lengths of these components indicates that the car's momentum was 30,000 kg·m/s and the pickup's was 40,000 kg·m/s. Because the mass of the car is 1000 kg, its velocity was

$$v_c = \frac{p_c}{m_c} = \frac{30{,}000\ \text{kg}\cdot\text{m/s}}{1000\ \text{kg}} = 30\ \text{m/s}$$

A similar calculation shows that the velocity of the pickup was 20 m/s. If the speed limit was 25 m/s, the car was speeding.

Example 6.4.2

Three children are standing together on a frictionless pond. At the count of three, they all push against each other in an attempt to escape to the edge of the pond. Child 1 (m_1 = 50 kg) moves north at 4 m/s. Child 2 (m_2 = 30 kg) moves east at 5 m/s. Child 3 (m_3 = 50 kg) moves at an angle θ south of west, as shown in the figure. Find the final velocity of Child 3.

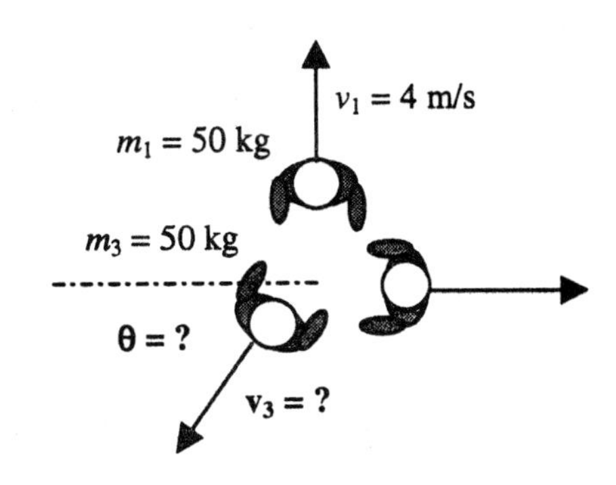

The initial total momentum of the three children is zero. Their total momentum must remain zero as there is no net external force acting on the children. Because momentum is a vector quantity, the southward component of child 3's momentum must be equal to the northward momentum of child 1.

$$p_1 = m_1 v_1 = (50 \text{ kg})(4 \text{ m/s}) = 200 \text{ kg} \cdot \text{m/s}$$

Likewise, the westward component of child 3's momentum must equal the eastward momentum of child 2.

$$p_2 = m_2 v_2 = (30 \text{ kg})(5 \text{ m/s}) = 150 \text{ kg} \cdot \text{m/s}$$

Therefore, the magnitude of child 3's momentum can be calculated using the Pythagorean theorem:

$$p_3 = \sqrt{p_1^2 + p_2^2} = \sqrt{(200 \text{ kg} \cdot \text{m/s})^2 + (150 \text{ kg} \cdot \text{m/s})^2} = 250 \text{ kg} \cdot \text{m/s}$$

Child #3's speed is therefore:

$$v_3 = \frac{p_3}{m_3} = \frac{250 \text{ kg} \cdot \text{m/s}}{50 \text{ kg}} = 5 \text{ m/s}$$

The direction can be found from a scale drawing with a protractor or using a calculator as explained in Section 3.1.

$$\tan\theta = \frac{200 \text{ kg} \cdot \text{m/s}}{150 \text{ kg} \cdot \text{m/s}} \quad \Rightarrow \theta = 53^\circ \text{ south of west}$$

Practice: Which child experiences the largest impulse?
Answer: The impulse each child experiences is equal to the child's change in momentum. They all started out at rest; that is, with zero momentum. Child 3 has the largest final momentum, so child 3 must have received the largest impulse.

Problems

1. A 2000-kg car is traveling at 20 m/s when the driver takes her foot off the gas pedal and slows to 10 m/s. What are the change in momentum and impulse experienced by the car?
2. An 8000-kg truck speeds up from 25 m/s to 30 m/s to pass a snowplow. What are the change in momentum and the impulse experienced by the truck?
3. A pitcher throws a baseball (m = 145 g) at 40 m/s toward home plate. The batter hits it straight back along its original path at 60 m/s. What is the impulse on the baseball?

4. A bowler with slippery fingers drops a 5-kg bowling ball. The ball is traveling vertically at 4 m/s just before striking the floor. It bounces straight back up at 1 m/s. What is the impulse on the bowling ball?

5. A 1600-kg car is traveling at 20 m/s when the driver takes her foot off the gas pedal. If we assume that the frictional forces are constant at 800 N, how long will it take the car to stop?

6. If the horizontal force of the road on the tires of a 1400-kg car is a constant 2000 N, how long will it take for the car to reach 30 m/s starting from rest?

7. Let's assume that the collision of a car with a wall takes 0.2 s. If the driver has a mass of 70 kg and is traveling at 24 m/s, what is the average force exerted on the driver by the seat belt? How does this compare to the driver's weight?

8. If the driver in the previous problem uses an air bag to increase the collision time to 0.5 s, what is the average force?

*9. A car traveling at 30 m/s crashes, causing air bags to inflate in front of the passengers. What minimum collision time must be provided by the air bags in order for the average forces on the passengers to remain less than their weights?

*10. Assume that a man with a mass of 80 kg jumps from an airplane without a parachute and hits a snow bank with a terminal speed of 50 m/s. What collision time would be needed for him to experience an average net force no larger than ten times his weight?

11. A pitcher throws a baseball (m = 145 g) at 40 m/s toward home plate. The batter hits it straight back along its original path at 50 m/s. If the ball is in contact with the bat for 0.05 s, what is the average force of the bat on the ball?

12. A hapless bowler with slippery fingers drops a 5-kg bowling ball on the floor. The ball is traveling vertically at 4 m/s just before striking the floor. It bounces straight back up at 1 m/s. If the collision with the floor lasts 0.1 s, what is the average force that the floor exerts on the ball?

13. A toy cannon with a mass of 4 kg fires a 0.2-kg ball with a horizontal velocity of 3 m/s. What is the recoil speed of the cannon?

14. A 50-kg boy standing on roller blades throws a 0.5-kg ball with a horizontal speed of 25 m/s. What is the recoil speed of the boy?

15. A man with a mass of 80 kg running at a speed of 5 m/s jumps onto a stationary skateboard with a mass of 4 kg. What is their combined speed?

16. A woman with a mass of 50 kg is riding on a giant 10-kg skateboard with speed of 5 m/s. If she jumps off the board in the backward direction with a speed of 2 m/s *relative to the skateboard*, what is the final speed of the board?

17. A ball traveling to the right at 2 m/s strikes an identical ball, which is initially at rest. After the collision, the second ball is moving to the right at 3 m/s. What is the final velocity (magnitude and direction) of the first ball?

18. In reporting the results of an experiment your classmate claims that a red ball traveling to the right at 4 m/s struck a blue ball of the same mass, which was initially stationary. After the collision, the red ball was observed moving to the right at 3 m/s. Why is this observation impossible?

19. A 2-kg ball traveling to the right with a speed of 4 m/s collides with a 4-kg ball traveling to the left with a speed of 2 m/s. After the collision the 2-kg ball travels to the left at 2 m/s. What is the velocity of the 4-kg ball after the collision?

20. A 2-kg ball traveling to the right with a speed of 6 m/s collides with a 4-kg ball traveling to the left with a speed of 4 m/s. If the 2-kg ball recoils to the left at 1 m/s, what

is the velocity of the 4-kg ball after the collision?

21. A 1200-kg car traveling west at 25 m/s collides with and sticks to a 1600-kg car traveling south at 35 m/s. What is the final speed of the two cars?

22. A 1000-kg car traveling north at 30 m/s collides with a 1500-kg car traveling west at 20 m/s. If they stick together, what is the final momentum (magnitude and direction) of the cars?

*23. A 1000-kg car traveling north at 30 m/s collides with a 1500-kg car traveling east. If the cars lock bumpers and travel directly northeast after the collision, how fast was the 1400-kg car traveling?

24. A 50-kg football player is running north at 6 m/s when he tackles a 100-kg player who is initially running east at 4 m/s. They slide together in the mud for 2 s before coming to rest. What is the magnitude of their combined momentum right before the collision? What is the average horizontal force by the mud on the players while sliding?

7—ENERGY

7.1 Energy of Motion

Example 7.1.1

How does the kinetic energy of a bullet fired from a rifle compare to the kinetic energy of a baseball thrown by a major league pitcher?

Before we can begin to answer this question, we need to have some additional data. The mass of a 30-06 bullet is about 0.01 kg and the muzzle velocity is about 900 m/s (2000 mph). An official baseball must have a mass within a few grams of 145 g. Finally, a major league pitcher can throw the ball up to 45 m/s (100 mph).

Let's begin by calculating the kinetic energy of the bullet.

$$KE_b = \tfrac{1}{2}m_b v_b^2 = \tfrac{1}{2}(0.01\text{ kg})(900\text{ m/s})^2 = 4050\text{ J}$$

The kinetic energy of the baseball is

$$KE_B = \tfrac{1}{2}m_B v_B^2 = \tfrac{1}{2}(0.145\text{ kg})(45\text{ m/s})^2 = 147\text{ J}$$

To compare the two kinetic energies, we divide them.

$$\frac{KE_b}{KE_B} = \frac{4050\text{ J}}{147\text{ J}} = 27.6$$

Therefore, the bullet has more than 27 times the kinetic energy of the baseball. We can also compare the two kinetic energies directly without calculating each one separately. This often has the advantage that the units cancel as long as each type of quantity is measured in the same units.

$$\frac{KE_b}{KE_B} = \frac{\tfrac{1}{2}m_b v_b^2}{\tfrac{1}{2}m_B v_B^2} = \left(\frac{m_b}{m_B}\right)\left(\frac{v_b}{v_B}\right)^2 = \left(\frac{10\text{ g}}{145\text{ g}}\right)\left(\frac{2000\text{ mph}}{100\text{ mph}}\right)^2 = 27.6$$

Practice: What is the ratio of the kinetic energies if the bullet has half the speed?
Answer: 6.9

7.2 Conservation of Kinetic Energy

In Section 6.2 we looked at momentum conservation in a collision of an air cart with an identical, stationary air cart. Let's look at the same three possibilities for the motion of the air carts after the collision to see if kinetic energy is conserved.

Example 7.2.1

In a collision of two air carts on a frictionless, horizontal air track, the carts have equal masses (m = 0.2 kg) and one of the carts is initially stationary. The moving cart has a velocity to the right (v = 0.5 m/s). What is the total kinetic energy of the two carts?

The total kinetic energy of the two cars is just the sum of the individual initial kinetic energies.

$$KE_i = \tfrac{1}{2}m_1v_1^2 + \tfrac{1}{2}m_2v_2^2 = \tfrac{1}{2}mv^2 + 0 = \tfrac{1}{2}(0.2\text{ kg})(0.5\text{ m/s})^2 = 0.025\text{ J}$$

Is kinetic energy conserved if the first cart stops and the second cart leaves with velocity v?

The total final kinetic energy in this case is

$$KE_f = \tfrac{1}{2}m_1v_1^2 + \tfrac{1}{2}m_2v_2^2 = 0 + \tfrac{1}{2}mv^2 = \tfrac{1}{2}(0.2\text{ kg})(0.5\text{ m/s})^2 = 0.025\text{ J}$$

Kinetic energy is conserved in this collision.

Is kinetic energy conserved if both carts leave the collision with one-half the original velocity?

$$KE_f = \tfrac{1}{2}m_1v_1^2 + \tfrac{1}{2}m_2v_2^2 = \tfrac{1}{2}m\left(\tfrac{1}{2}v\right)^2 + \tfrac{1}{2}m\left(\tfrac{1}{2}v\right)^2 = \tfrac{1}{4}mv^2$$
$$= \tfrac{1}{2}(0.2\text{ kg})(0.25\text{ m/s})^2 + \tfrac{1}{2}(0.2\text{ kg})(0.25\text{ m/s})^2 = 0.0125\text{ J}$$

Therefore, kinetic energy is not conserved for this possibility. This is a totally inelastic collision as the two carts have the same velocity after the collision.

Is kinetic energy conserved if the cart that was initially moving rebounds in the opposite direction with $\tfrac{1}{2}v$ and the struck cart leaves with $\tfrac{3}{2}v$?

$$KE_f = \tfrac{1}{2}m_1v_1^2 + \tfrac{1}{2}m_2v_2^2 = \tfrac{1}{2}m\left(\tfrac{3}{2}v\right)^2 + \tfrac{1}{2}m\left(\tfrac{1}{2}v\right)^2 = \tfrac{5}{4}mv^2$$
$$= \tfrac{1}{2}(0.2\text{ kg})(0.25\text{ m/s})^2 + \tfrac{1}{2}(0.2\text{ kg})(0.75\text{ m/s})^2 = 0.0625\text{ J}$$

Notice that the final kinetic energy for this case is larger than the initial kinetic energy. Conservation of energy forbids this from happening unless there is a source for the extra energy. This could happen if one of the carts held a compressed spring that was released during the collision. The spring would not affect the conservation of momentum because it is an internal force.

As discussed above, not all collisions are elastic. Any collision in which the objects stick together after the collision is totally inelastic and the maximum amount of kinetic energy that is allowed by the conservation of momentum is lost. However, many collisions are intermediate in that some kinetic energy is lost, but not the maximum amount.

Example 7.2.2

A 0.5-kg ball falls from a height of 45 m, hits the ground, and bounces to a height of 20 m. What fraction of its kinetic energy is lost in the collision with the ground?

According to the table in Section 2.7, a ball falling from a height of 45 m takes 3 s to hit the ground and is traveling with a speed of 30 m/s. The same table shows that ball falling from 20 m will hit the ground at 20 m/s after 2 s. Because free fall is symmetric, the ball must have left the ground with a speed of 20 m/s. This information allows us to calculate the two kinetic energies.

$$KE_i = \tfrac{1}{2} m v_i^2 = \tfrac{1}{2}(0.5\text{ kg})(30\text{ m/s})^2 = 225\text{ J}$$

$$KE_f = \tfrac{1}{2} m v_f^2 = \tfrac{1}{2}(0.5\text{ kg})(20\text{ m/s})^2 = 100\text{ J}$$

Therefore, the fractional change in kinetic energy is

$$\frac{KE_f - KE_i}{KE_i} = \frac{100\text{ J} - 225\text{ J}}{225\text{ J}} = -0.556 = -55.6\,\%$$

The fractional change can be calculated more directly without calculating the individual kinetic energies first.

$$\frac{KE_f - KE_i}{KE_i} = \frac{\tfrac{1}{2}mv_f^2 - \tfrac{1}{2}mv_i^2}{\tfrac{1}{2}mv_i^2} = \frac{v_f^2 - v_i^2}{v_i^2} = \left(\frac{v_f}{v_i}\right)^2 - 1$$

$$= \left(\frac{20\text{ m/s}}{30\text{ m/s}}\right)^2 - 1 = 0.444 - 1 = -0.556$$

This last relationship has the nice feature of showing us that the fractional change in kinetic energy does not depend on the mass of the ball.

Practice: What is the fractional loss in kinetic energy if the ball rebounds to a height of 5 m?
Answer: 88.9%

7.3 Changing Kinetic Energy

We can show how work changes the kinetic energy of an object by using Newton's second law to write the force as the product of the mass and the acceleration.

$$W = Fd = mad$$

The kinetic energy depends on the object's speed, so we modify this expression by making two substitutions that eventually get speed into the equation. In Section 2.8 we found that under the action of a constant force, an object starting from rest moves a distance given by

$$d = \tfrac{1}{2}at^2$$

Using this relation, our expression becomes

$$W = mad = ma\left(\tfrac{1}{2}at^2\right) = \tfrac{1}{2}m\left(at\right)^2$$

We also learned in Chapter 2 that

$$v_f = v_i + at$$

Because v_i is zero in this case, $v_f = at$. Therefore, we can replace the term in parentheses by v_f to obtain

$$W = \tfrac{1}{2}mv_f^2$$

This equation says that the work done on the object is equal to the kinetic energy it acquires. Because the object started with zero kinetic energy, this equation also represents the change in the kinetic energy due to the work done on the object. If the object had been given an initial velocity, the substitutions would have been a little more difficult, but the results would have been the same; the work done is equal to the change in the kinetic energy.

$$W = \Delta KE = \tfrac{1}{2}mv_f^2 - \tfrac{1}{2}mv_i^2 = \tfrac{1}{2}m\left(v_f^2 - v_i^2\right)$$

Example 7.3.1

How much work is required to accelerate a 1600-kg car from rest to a speed of 30 m/s?

Because the work is just equal to the change in kinetic energy,

$$W = \tfrac{1}{2}mv_f^2 - 0 = \tfrac{1}{2}(1600\text{ kg})(30\text{ m/s})^2 = 7.2 \times 10^5\text{ J}$$

Practice: How much work is required to accelerate the car to one-half this speed?
Answer: 1.8×10^5 J

Example 7.3.2

How much work is required to accelerate a 1600-kg car from an initial speed of 15 m/s to a final speed of 30 m/s?

The work is equal to the change in kinetic energy.

$$W = \tfrac{1}{2}m\left(v_f^2 - v_i^2\right) = \tfrac{1}{2}(1600\text{ kg})\left((30\text{ m/s})^2 - (15\text{ m/s})^2\right) = 5.4 \times 10^5\text{ J}$$

Example 7.3.3

What average force is required to accelerate a 6000-kg truck from rest to a final speed of 20 m/s over a distance of 1.5 km?

We can equate the work to the change in kinetic energy (the final kinetic energy in this case) and solve for the average force.

$$Fd = \Delta KE = \tfrac{1}{2}mv_f^2$$

$$F = \frac{\tfrac{1}{2}mv_f^2}{d} = \frac{\tfrac{1}{2}(6000\text{ kg})(20\text{ m/s})^2}{1.5 \times 10^3\text{ m}} = 800\text{ N}$$

Practice: What average force would be required for the truck to reach the same final speed in 0.5 km?
Answer: 2400 N

7.4 Gravitational Potential Energy

Example 7.4.1

How much gravitational potential energy does a 70-kg mountain climber gain in walking from sea level to the top of a 3000-m peak?

$$GPE = mg\Delta h = (70\text{ kg})(9.8\text{ m/s}^2)(3000\text{ m}) = 2.06 \times 10^6\text{ J}$$

Notice that we did not need to specify how far the climber actually walked from the starting place to the top of the mountain. The change in gravitational potential energy depends only on the change in vertical height.

Practice: What is the gain in potential energy if the mountain climber starts at an elevation of 1000 m?
Answer: 1.37×10^6 J

7.5 Conservation of Mechanical Energy

Example 7.5.1

How fast does a 0.4-kg ball hit the ground when it is dropped from a height of 50 m?

Conservation of mechanical energy tells us to set the final value of the mechanical energy equal to its initial value.

$$KE_f + GPE_f = KE_i + GPE_i$$

Let's choose the zero value for the gravitational potential energy to be ground level. Noting that the initial kinetic energy is zero, we have

$$KE_f = GPE_i$$

We can now substitute in the formulas for the kinetic energy and gravitational potential energy.

$$\tfrac{1}{2}mv_f^2 = mgh_i$$

Canceling the mass m and solving for v_f^2, we get our answer.

$$v_f^2 = 2gh_i = 2\left(9.8\ \text{m/s}^2\right)(50\ \text{m}) = 980\ \text{m}^2/\text{s}^2$$

$$v_f = 31.3\ \text{m/s}$$

We know that this answer is reasonable because the ball must be in the air for 3.19 s to obtain this speed and a ball will fall 44.1 m from rest in 3 s. Notice also that the mass does not affect the final speed. This is what we expect for free fall.

Practice: What is the speed if the ball is dropped from twice the height?
Answer: 44.3 m/s

Example 7.5.2

A ball is shot from the top of a 50-m high cliff with a speed of 5 m/s. With what speed does it hit the plain below?

We set the gravitational potential energy equal to zero at the bottom of the cliff and equate the final value of the mechanical energy to its initial value.

$$\tfrac{1}{2}mv_f^2 + 0 = \tfrac{1}{2}mv_i^2 + mgh_i$$

Solving for v_f^2 and plugging in the given values yields

$$v_f^2 = v_i^2 + 2gh_i = (5\text{ m/s})^2 + 2(9.8\text{ m/s}^2)(50\text{ m}) = 1000\text{ m}^2/\text{s}^2$$

$$v_f = 31.6\text{ m/s}$$

Note that we did not need to know the direction of the ball to compute the final speed. The ball could have been launched straight up, at an angle upward, horizontal, or downward without affecting the result.

Practice: Show that you get the same answer by choosing the gravitational potential energy to be zero at the top of the cliff.

Example 7.5.3

You swing on a rope from a ledge that is 5 m above a lake. Your 5-m rope is fastened to a tree limb that is 6 m above the lake. When you jump, you swing past the bottom of the swing and rise to a height of 2 m above the water before you let go. How fast will you be traveling just before you hit the water?

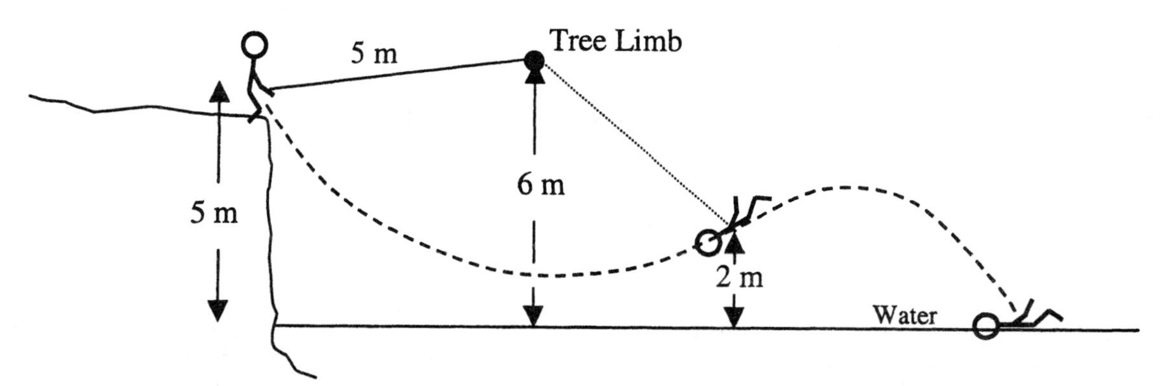

First, we note that the kinematic equations are not applicable during the portion of the motion when you are swinging on the rope because your acceleration is continually changing both magnitude and direction. We solve this problem using conservation of mechanical energy because the rope doesn't do any work as the tension is always perpendicular to your path. Your kinetic energy just before you hit the water must be the same as your potential energy when you were standing on the ledge.

$$mgh_{initial} = \tfrac{1}{2}mv_{final}^2$$

$$v_{final} = \sqrt{2gh_{initial}} = \sqrt{2(10\text{ m/s}^2)(5\text{ m})} = 10\text{ m/s}$$

Notice that this result does not depend on where you let go of the rope. Indeed, we do not need to worry about any of the details of your swing. You could have included a double flip before hitting the water and the final speed would be the same. It is important to recognize that with the simplicity of this approach comes a price. We do not learn anything about *where* you land or *when* you land. We also lose any information about the *direction* you are traveling when you land.

7.6 Power

Example 7.6.1

What power is required to accelerate a 1600-kg car from rest to a speed of 30 m/s in a time of 6 s?

We already calculated in Example 7.3.1 that the car gained 7.2×10^5 J of kinetic energy. Therefore, the average power required is

$$P = \frac{\Delta E}{\Delta t} = \frac{7.2 \times 10^5 \text{ J}}{6 \text{ s}} = 1.2 \times 10^5 \text{ W} = 120 \text{ kW}$$

Practice: What average power would be required in Example 7.3.2 if the change in speed takes place in 3 s?
Answer: 240 kW

Problems

1. What is the kinetic energy of a 50-kg diver if she hits the water at 4 m/s?
2. If an 18-wheeler has a mass of 30,000 kg and a speed of 30 m/s, what is its kinetic energy?
3. What is the speed of a 1000-kg sports car with a kinetic energy of 400,000 J?
4. If an 80-kg jogger has a kinetic energy of 600 J, how fast is the jogger running?
5. A 1-kg air-hockey puck moving at 8 m/s collides head-on with a 2-kg puck traveling at 4 m/s in the opposite direction. The pucks rebound in opposite directions with the 1-kg puck traveling at 6 m/s and the 2-kg puck at 3 m/s. Are momentum and kinetic energy conserved?
6. A 1-kg air-hockey puck moving at 6 m/s collides head-on with a stationary 2-kg puck. The 1-kg puck recoils in the backward direction with a speed of 2 m/s while the 2-kg puck moves in the forward direction with a speed of 4 m/s. Are momentum and kinetic energy conserved in this collision?
7. A block of mass m moving to the right with speed v_0 collides with a block of mass $4m$, which is initially stationary. The small block rebounds to the left at speed $\frac{1}{5}v_0$ while the large block moves to the right with speed $\frac{3}{10}v_0$. Show that momentum is conserved and find how much kinetic energy has been lost in the collision.

8. A block of mass m moving to the right with speed v_0 collides with a block of mass $3m$, which is initially stationary. The two blocks stick together and move away with speed $\frac{1}{4}v_0$. Show that momentum is conserved and find how much kinetic energy has been lost in the collision.

*9. A 1-kg air-hockey puck moving at 6 m/s collides head-on with a stationary 2-kg puck. If the two pucks stick together, what is their final kinetic energy?

*10. A 10-g bullet moving horizontally at 900 m/s embeds itself in a 2-kg wooden block, which is initially at rest. At what speed does the block-bullet system move immediately after the collision? What fraction of the initial kinetic energy was lost during the collision?

11. What average force would it take to catch a 0.5-kg ball with a speed of 40 m/s if your hands give a distance of 20 cm?

12. What average force is required for a sprinter with a mass of 75 kg to reach a speed of 10 m/s over a distance of 15 m?

13. A 0.5-kg air-hockey puck has an initial kinetic energy of 0.4 J. What will its final kinetic energy be after a force of 0.4 N acts on it for a distance of 0.2 m?

14. A force of 0.2 N acts on a 0.4-kg air-hockey puck for a distance of 0.6 m. What is the final kinetic energy of the puck if it had an initial kinetic energy of 0.2 J?

15. Use the data in the feature "Stopping Distances for Cars" in the text to calculate the braking distance for a car traveling at 35 mph.

16. Use the data in the feature "Stopping Distances for Cars" in the text to calculate the braking distance for a car traveling at 45 mph.

17. A bucket of water is raised from a well using a rope wrapped around a cylinder. The cylinder is turned by a crank that is 48 cm long. The handle follows a circle that has a circumference of 3 m. If the force on the crank is 100 N, how much work is done each time it goes around?

18. The handle of a grinding wheel goes around a circle with a circumference of 1 m. If it requires a force of 150 N to turn the handle at a steady rate, how much work is performed each revolution?

19. Block A slides across a frictionless table top and collides elastically with an identical block B, which is initially at rest. After the collision, block A is a rest and block B continues along the surface and up a frictionless ramp to a height of 40 cm above the tabletop. If the two blocks had instead stuck together, how high would they travel up the ramp?

20. A 10-kg block slides from rest down a frictionless ramp such that the center of the block drops a distance of 0.8 m. After the block has reached the bottom of the ramp, it encounters a 2-m section of rough track that creates 30-N frictional force on the block. (a) What is the block's kinetic energy just before it hits the rough section? (b) What is the block's kinetic energy after the rough section? (c) What is the block's speed after the rough section?

21. A man with a mass of 80 kg falls 8 m. How much gravitational potential energy does he lose?

22. A ball that weighs 2 N can be placed on a counter that is 1 m above the floor or on a shelf that is 2 m above the floor. (a) If we chose the zero value of gravitational potential energy to be at the floor, what is the value of the gravitational potential energy when it is on the counter? On the shelf? (b) How much does the gravitational potential energy change when the ball is moved from the counter to the shelf? (c) Do your answers to these questions change if the gravitational potential energy is chosen to be zero on the counter?

23. Show that the units of kinetic energy, work, and gravitational potential energy are the same.

24. Show that the kinetic energy of an object can be written $KE = p^2/2m$, where p is the object's momentum and m is its mass.

25. A 30-kg child slides down a frictionless slide that is 2 m tall. What is the child's kinetic energy at the bottom of the slide?

26. A man with a mass of 70 kg falls 10 m. How much kinetic energy does he gain?

27. What is the speed of a man who falls 9.8 m?

28. What is the speed of the child in Problem 25 at the bottom of the slide?

29. A pendulum bob changes height by a distance of 50 cm from one end of its swing to its lowest point. What is the speed of the pendulum at the lowest point?

30. If the pendulum in Problem 29 has a length of 3 m and a bob of mass 2 kg, what is the tension in the string when the pendulum is at its lowest point?

31. A ball is hit vertically upward with a speed of 40 m/s. How high will it go?

*32. A model rocket weighs 1 N and develops an average thrust of 5 N for 2 s. At the end of this time, the rocket has obtained a height of 80 m and a speed of 80 m/s. (Neglect the loss in mass due to the expelled gases.) (a) How much work was done by the net force on the rocket? (b) Calculate the value of the kinetic energy of the rocket using the formula $\frac{1}{2}mv^2$. (c) Should the answers to parts (a) and (b) be the same? Explain. (d) How much higher will the rocket coast before this kinetic energy is converted to gravitational potential energy?

33. What average power is required for a runner with a mass of 70 kg to reach a speed of 9 m/s during a time of 2 s?

34. What average power is dissipated by the brakes of a 1800-kg car if comes to a stop from a speed of 30 m/s in 3 s?

35. What happened to the work in Problem 17 if the bucket is raised at a steady rate?

36. What happens to the work in Problem 18?

37. Show that the product of force and velocity has the units of power, not energy.

*38. A 1982 Lincoln Continental requires 16.5 hp (12.3 kW) to maintain a speed of 50 mph (22.4 m/s) on a level highway. This power is used to overcome friction and air resistance. What is the force required to maintain this speed? (See the previous problem for a hint.)

8 — ROTATION

8.1 Torque

Example 8.1.1

Where do the two girls shown in Figure 8-6 in the text need to sit to balance the seesaw?

We can calculate the required location of the heavier child if we know the weights of both children and the location of the smaller child. Suppose the children weigh 300 N and 400 N, and that the lighter child is 2 m from the pivot point. The lighter child produces a torque of

$$\tau = Fr = (300\text{ N})(2\text{ m}) = 600\text{ N}\cdot\text{m}$$

The torque created by the heavier child must have the same value for the seesaw to be balanced. Rearranging the torque equation in order to solve for the moment arm yields

$$r = \frac{\tau}{F} = \frac{600\text{ N}\cdot\text{m}}{400\text{ N}} = 1.5\text{ m}$$

The heavier child should sit 1.5 m from the pivot.

Practice: Where should the heavier child sit if the lighter child has a weight of only 200 N?
Answer: 1 m from the pivot

Example 8.1.2

A 100-g mass hangs from the 70-cm mark of a meter-stick balance. Where would you need to hang a 50-g mass to balance the system?

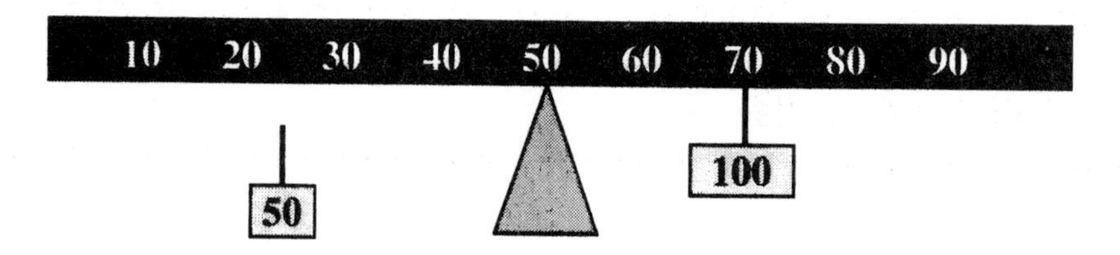

The torque on the left-hand side (side 1) must be equal to the torque on the right-hand side (side 2) for the system to be balanced.

$$m_1 g r_1 = m_2 g r_2$$

$$r_1 = \frac{m_2}{m_1} r_2 = \frac{100\text{ g}}{50\text{ g}}(20\text{ cm}) = 40\text{ cm}$$

Notice that we did not use the position (70 cm) of the 100-g mass; we used the distance of the mass from the pivot point. Likewise, our answer tells us how far from the pivot point to put the 50-g mass from the pivot point. Therefore, we should place the 50-g mass at the 10-cm mark.

Practice: If a 40-g mass hangs at the 80-cm mark, where would you place a 100-g mass to balance the system?
Answer: At the 38-cm mark.

Example 8.1.3

A 50-g mass hangs from the 70-cm mark and a 40-g mass hangs from the 90-cm mark of a meter-stick balance. Where would you need to hang a 100-g mass to balance the system?

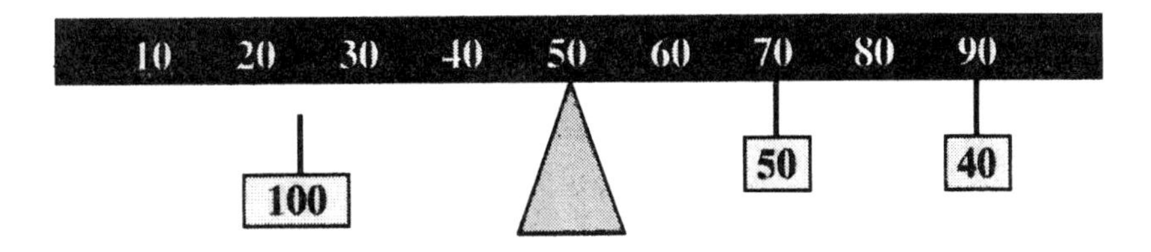

The torque on the left-hand side must be equal to the torque on the right-hand side for the system to be balanced.

$$m_1 g r_1 = m_2 g r_2 + m_3 g r_3$$

$$r_1 = \frac{m_2 r_2 + m_3 r_3}{m_1} = \frac{(50\text{ g})(20\text{ cm}) + (40\text{ g})(40\text{ cm})}{100\text{ g}} = 26\text{ cm}$$

Therefore, we should place the 100-g mass at the 24-cm mark.

Practice: If a 40-g mass hangs at the 60-cm mark and a 30-g mass hangs at the 80-cm mark, where would you place a 50-g mass to balance the system?
Answer: At the 12-cm mark.

8.2 Rotational Inertia

In the text we stated that the rotational inertia of a body depended on the distribution of mass as well as the total mass of the object. The expression for the rotational inertia I of a point mass m moving in a circle of radius r is

$$I = mr^2 \qquad \textit{rotational inertia}$$

Notice that the rotational inertia depends on the square of the radius. This means that the mass farthest from the axle contributes the most to the rotational inertia. The rotational inertia of an extended object is found by summing the contributions of all the "particles" that make up its bulk.

Example 8.2.1

What is the rotational inertia of a thin ring of radius *a* and total mass *M* rotating about an axle through the center of the ring and perpendicular to the plane of the ring?

We need to sum up the contributions of each little piece of mass. Let's imagine that we break the ring up into small pieces, each with a mass of m. Then our sum becomes

$$I = ma^2 + ma^2 + ma^2 + \ldots..$$

Factoring out the common factor a^2, we have

$$I = a^2 (m + m + m + \ldots..)$$

Because the sum in parentheses is just the total mass M of the ring, we find that

$$I = Ma^2$$

If $a = 10$ cm and $M = 1$ kg, we find that the rotational inertia is

$$I = Ma^2 = (1\,\text{kg})(0.1\,\text{m})^2 = 0.01\,\text{kg} \cdot \text{m}^2$$

Practice: What is the value for the rotational inertia if the ring has twice the radius and the same mass?
Answer: 0.04 kg·m^2

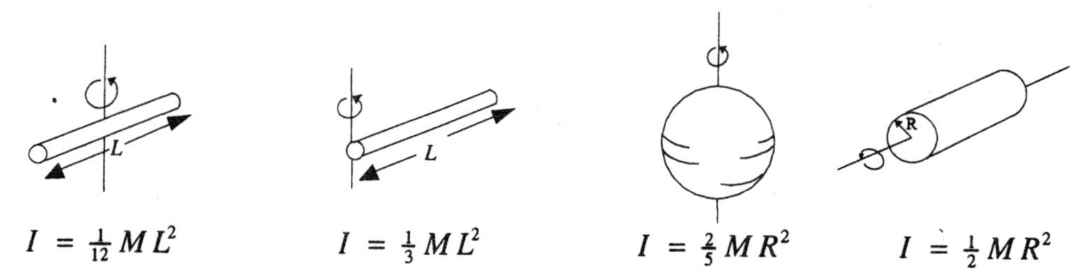

Fig. 8.2.1

Figure 8.2.1 shows expressions for the rotational inertia of some symmetric objects. Notice that in two cases the same object rotates about different axes and has different rotational inertias. An object can have many different values for its rotational inertia but only one for its inertial mass.

Newton's second law for rotation can be written

$$\tau = I\alpha$$

where τ is the torque, I is the rotational inertia, and α is the rotational acceleration. This particular form of the second law is a direct analog of the translational equation. However, it does require that the angles be measured in a special unit known as a radian.

$$1 \text{ radian} = \frac{360°}{2\pi} = 57.3°$$

Example 8.2.2

What torque would be required to rotate the ring in the previous example with an angular acceleration of 1 rev/min^2?

We first need to convert the angular acceleration into the proper units of rad/s. Note that the definition of a radian tells us that $360° = 2\pi$ radians = 1 rev.

$$\alpha = \frac{1 \text{ rev}}{(1 \text{ min})^2}\left[\frac{6.28 \text{ rad}}{1 \text{ rev}}\right]\left[\frac{1 \text{ min}}{60 \text{ s}}\right]^2 = 0.00174 \text{ rad/s}^2$$

The torque is equal to the angular acceleration multiplied by the rotational inertia we calculated in the previous example.

$$\tau = I\alpha = \left(0.01 \text{ kg}\cdot\text{m}^2\right)\left(1.74 \times 10^{-3} \text{ rad/s}^2\right) = 1.74 \times 10^{-5} \text{ N}\cdot\text{m}$$

8.3 Conservation of Angular Momentum

The direction of the angular momentum vector is not intuitively obvious. The first vectors that we discussed seemed reasonable because in each case the direction of the vector was obvious. A force to the right is represented by a vector pointing to the right; the velocity vector points in the direction that the object is moving. The direction of the acceleration vector is not as intuitive as it doesn't usually point in the direction the particle is moving, but it always points in the direction of the net force. In uniform circular motion, for example, the acceleration points toward the center of the circle—perpendicular to the velocity vector. In projectile motion, the acceleration always points vertically downward.

Let's look at something that has a constant angular momentum to see how a direction might be associated with the angular momentum vector. Consider a ball whirling in a circle on the end of a string. If the speed of the ball is constant and the length of the string stays the same, the angular momentum is constant. The directions of the velocity and acceleration vectors are continually changing and therefore cannot be used to describe a constant angular momentum. However, there is a direction that is constant. This is the line that passes through the center of the circular path perpendicular to the plane of the circle.

There is still a problem of uniqueness. The vector can be aligned along this axis but point in one of two directions. A vector must be unique to describe a particular motion. If it doesn't, it is of little value. The specific choice is arbitrary; the convention is to curl the fingers of your *right hand* along the direction of motion. Your thumb then points along the axis in the direction of the angular momentum as shown in Figure 8-2 in the text.

Conservation of angular momentum requires that the direction of the angular momentum, as well as its size, remain constant. In many cases, the direction remains constant and we need only work with the magnitude.

Example 8.3.1

The Earth is closest to the Sun in January (147 million km) and farthest in July (152 million km). The Earth has a known speed of 28.8 km/s when it is farthest from the Sun. What is its speed when it is closest?

Because the force on the Earth is always directed toward the Sun, there is no torque on Earth and its angular momentum is conserved. At the locations on the orbit when the Earth is closest to and farthest from the Sun, the velocity is perpendicular to the radius and the calculation of the angular momentum is easy. Using the subscripts f for farthest and n for nearest, we equate the two angular momenta.

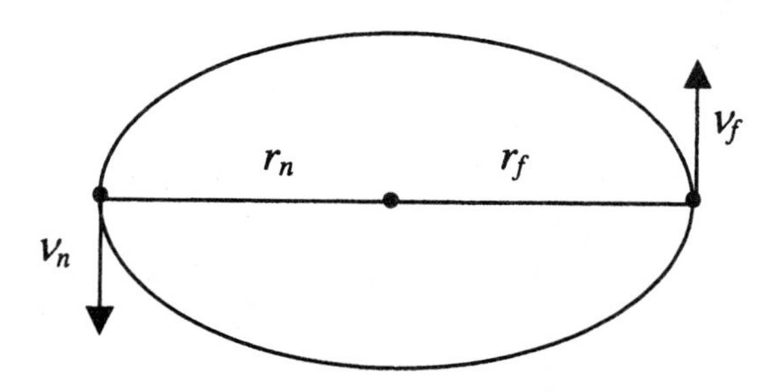

$$M_E r_n v_n = M_E r_f v_f$$

We can cancel the mass M_E of the Earth and solve the relationship for any one of the remaining quantities. Solving for v_n and substituting the given values for the others, we have

$$v_n = v_f \frac{r_f}{r_n} = (28.8 \text{ km/s})\left(\frac{1.52 \times 10^8 \text{ km}}{1.47 \times 10^8 \text{ km}}\right) = 29.8 \text{ km/s}$$

Example 8.3.2

If a diver can execute a somersault in the tuck position in 1 s, how long would it take in the layout position?

Let's assume that we use the values given in Figure 8-14 in the text. Then the relative values for the rotational inertia are 230 and 830 for the tuck and layout positions, respectively. Conservation of angular momentum requires that

$$I_\ell \omega_\ell = I_t \omega_t$$

where I_l and I_t are the rotational inertia for the layout and tuck positions, respectively, and ω_l and ω_t are the corresponding angular speeds in rad/s. Because $\omega = 2\pi/T$, we can also express this in terms of the time T for one revolution.

$$\frac{I_\ell}{T_\ell} = \frac{I_t}{T_t}$$

Solving for T_ℓ we obtain the time needed to complete the somersault in the layout position.

$$T_\ell = T_t \frac{I_\ell}{I_t} = (1\,\text{s})\left(\frac{830}{230}\right) = 3.61\,\text{s}$$

Practice: Given that the relative rotational inertia for the pike position is 340, how long would a somersault in the pike position require?
Answer: 1.48 s

Example 8.3.3

A hockey puck of mass M and radius R is rotating on a pond of frictionless ice at a rotational speed of 24 rad/s. A hollow metal cylinder (a tin can with the top and bottom removed) having the same radius and the same mass is dropped such that it lands centered on the spinning puck. The cylinder starts to rotate due to the kinetic frictional force between the puck and the cylinder, and soon both are spinning together at the same rate. What is this final rotational speed?

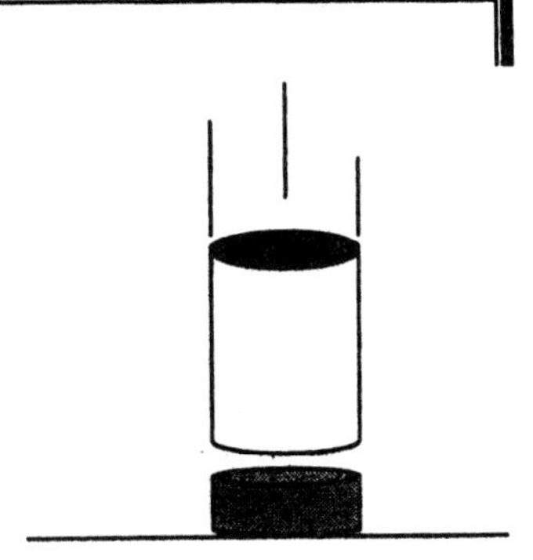

The rotational momentum of a system must be conserved during a collision as long as there are no external torques. The frictionless ice exerts no torque on the hockey puck, and the torques due to the frictional force between the cylinder and the puck are internal to the system, so the angular momentum of the cylinder and puck after the collision must be the same as the angular momentum of the puck before the collision.

$$I_{puck}\omega_{initial} = \left(I_{puck} + I_{cylinder}\right)\omega_{final}$$

$$\omega_{final} = \frac{I_{puck}}{I_{puck} + I_{cylinder}}\omega_{initial}$$

This result is valid for any puck and any cylinder. For our case, we have

$$I_{cylinder} = MR^2$$

$$I_{puck} = \tfrac{1}{2}MR^2$$

$$\Rightarrow I_{cylinder} = 2I_{puck}$$

Therefore,

$$\omega_{final} = \frac{I_{puck}}{I_{puck} + 2I_{puck}} \omega_{initial} = \tfrac{1}{3}(24 \text{ rad/s}) = 8 \text{ rad/s}$$

Practice: What is the final speed if an identical hockey puck is dropped from rest onto the spinning puck?
Answer: 12 rad/s

Problems

1. Two kids with masses of 50 kg and 60 kg want to balance a seesaw. If the lighter kid sits 1.5 m from the center, where should the heavier kid sit?

2. A 30-g mass is hung 35 cm from the pivot point of a balance. Where would you hang a 50-g mass to balance the system?

3. Two boys are sitting on a plank placed across a log. The heavier boy has a mass of 60 kg and sits 1.2 m from the center of the plank. If the other boy sits 1.6 m from the center, what is his mass?

4. A bag of candy is hung at the end of a meter stick that is suspended from its center. If a 100-g mass placed 38 cm from the center of the meter stick balances the bag of candy, what is the mass of the candy?

5. A 200-g mass is hanging from the 75-cm mark and a 100-g mass is hanging from the 65-cm mark of a meter-stick balance? Where would you place a 500-g mass to balance the system?

6. Two girls sit on one end of a seesaw. The first girl has a mass of 30 kg and sits 1.2 m from the pivot point. The second girl has a mass of 40 kg and sits 1.5 m from the pivot point. Where should a woman with a mass of 60 kg sit to balance the seasaw?

7. A 100-g mass is hanging from the 85-cm mark and a 75-g mass is hanging from the 0-cm mark of a meter-stick balance. Where would you place a 25-g mass to balance the system?

8. A man with a mass of 80 kg sits at one end of a 3-m long seasaw. A woman with a mass of 60 kg sits at the other end. Where should a 30-kg child sit to balance the seasaw?

9. If the rotational speed of a flywheel changes from 72 rev/s to 90 rev/s in a time of 30 s, what is its average rotational acceleration?

10. If a phonograph turntable takes 2 s to reach its rotational speed of 78 rpm, what is its average rotational acceleration?

11. If the wheel on a car accelerates from rest at a constant rate of 1.4 rev/s^2, how fast will it be turning after 12 s?

12. If a videodisc player can accelerate videodiscs at a rotational acceleration of 10 rev/s^2, how fast will it be rotating after 3 s?

*13. An amusement park ride accelerates from rest at a constant rate of 0.01 rev/s^2. How many times will it have rotated in 30 s?

*14. If a merry-go-round starts from rest and accelerates at a constant rate of 8 rev/min^2, how many times will it rotate in 5 min?

15. What is the rotational acceleration of a wheel if it has a torque of 600 N·m and a rotational inertia of 150 kg·m^2?

*16. A uniform, solid wheel has a radius of 1.5 m and a mass of 40 kg. A string is wrapped around the outside of the wheel and a 1-kg mass is hung from its end. What is the rotational acceleration of the wheel?

17. What is the rotational inertia of a thin ring about its center if it has a mass of 8 kg and a radius of 0.4 m?

18. A solid sphere of mass 22 kg and radius 0.3 m is rotated about an axis through its center. A solid rod of mass 3 kg and length 1 m is rotated about an axis through one of its ends. Given the same torque on both, which will experience the greater angular acceleration?

19. Halley's comet orbits the Sun in a huge elliptical path. Its greatest distance from the Sun is 60 times the smallest distance. If the comet has a speed of 193,000 km/h when it is nearest the Sun, what is its speed when it is farthest away?

20. The closest Mars gets to the Sun is 207 million km and its farthest distance away is 249 million km. If Mars' orbital speed at closest approach is 26.5 km/s, what is its orbital speed when it is farthest away?

21. A metal ring has a mass of 18 kg and a radius of 0.5 m. If it is rotating about its axis at 5 rpm, what is its angular momentum?

22. The Sun spins on its own axis with a period of 25 days, which gives a rotational speed of 2.91×10^{-6} rad/s. For this problem we will treat the Sun as a solid uniform sphere of radius 6.96×10^{8} m and mass 1.99×10^{30} kg. What is the Sun's angular momentum? (See Figure 8.2.1 for the rotational inertia.)

23. An ice skater is spinning with a rotational speed of 2 rev/s. When he extends his arms and one leg, his rotational inertia increases by a factor of three. What is his final rotational speed?

24. A springboard diver is executing a triple somersault She begins in the layout position with a rotational speed of 1 rev/s. When she changes to the tuck position, her moment of inertia decreases by a factor of 4. What is her rotational speed in the tuck position?

25. A solid cylindrical disk is rotating at 60 rev/min with its axle pointing vertically. An identical disk with no rotation is dropped directly onto the rotating disk. What is the final rotational speed of the two disks?

26. A merry-go-round with a radius of 1 m and a rotational inertia of 40 kg·m^2 is rotating at 2 rad/s. A boy with a mass of 40 kg is standing next to the merry-go-round. If he sits on the edge of the merry-go-round, what will the final rotational speed of the system be?

27. Show that the units of angular momentum are the same as those of torque multiplied by time.

28. Show that the units of momentum and impulse are the same.

9 — CLASSICAL RELATIVITY

9.1 Comparing Velocities

In the text we described a situation in which your friends were rolling a ball on the floor of a moving van. We stated that the velocity of the ball measured relative to the ground was equal to the *vector* sum of the velocity of the van measured relative to the ground and the velocity of the ball measured relative to the van. We can translate these words into a vector equation.

$$\boldsymbol{v}_{og} = \boldsymbol{v}_{os} + \boldsymbol{v}_{sg}$$

where the subscripts o, g, and s refer to the object, ground, and moving reference system, respectively. Therefore, $\boldsymbol{v}_{og}$ is the velocity of the object (the ball) measured relative to the ground, $\boldsymbol{v}_{os}$ is the velocity of the object measured relative to the moving reference system (the van), and $\boldsymbol{v}_{sg}$ is the velocity of the system measured relative to the ground. The key to finding velocities using this vector relationship is to carefully identify the meaning of each symbol and to remember their vector properties. In motion along a straight line, we can once again use plus and minus signs to denote the directions of the velocities.

Example 9.1.1

Let's work the example in the text using this equation. The ball rolled on the floor at 2 m/s toward the east. The van was traveling at 3 m/s toward the east. What is the velocity of the ball measured relative to the ground?

Choosing east as the positive direction and identifying the ball as the object and the van as the moving reference system, we have

$$v_{og} = v_{sg} + v_{os} = +2\ \text{m/s} + 3\ \text{m/s} = +5\ \text{m/s}$$

Therefore, the ball is moving 5 m/s eastward relative to the ground. If the ball were rolled toward the back of the van, we would have

$$v_{og} = v_{sg} + v_{os} = -2\ \text{m/s} + 3\ \text{m/s} = +1\ \text{m/s}$$

In this case the ball is still moving toward the east (indicated by the plus sign), but its speed relative to the ground has been reduced to 1 m/s.

Practice: What is the speed of the ball relative to the ground if it is rolled toward the back of the van at 4 m/s?
Answer: –1 m/s or 1 m/s westward

Example 9.1.2

What would the ball's speed be relative to the ground in Example 9.1.1 if it were rolled directly to the side of the van?

We can draw a vector diagram like that in Figure 9.1.1 and measure the size of the resultant vector. Alternatively, because the two velocity vectors are perpendicular to each other, we can also obtain the speed using the Pythagorean theorem.

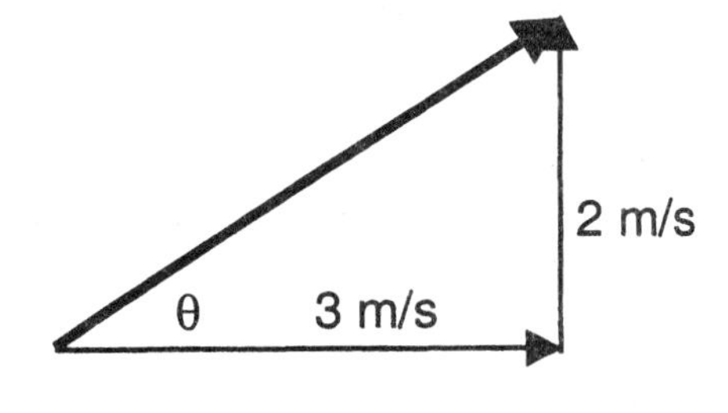

Fig 9.1.1

$$v_{og} = \sqrt{v_{os}^2 + v_{sg}^2} = \sqrt{(2\text{ m/s})^2 + (3\text{ m/s})^2}$$
$$= \sqrt{13\text{ m}^2/\text{s}^2} = 3.61\text{ m/s}$$

The direction θ of the velocity of the ball relative to the ground can be measured on the diagram to be 34°, or determined with your calculator as described in Section 3.1.

9.2 Realistic Inertial Forces

Let's use the elevator described in the text as our example situation for solving problems involving accelerating reference systems. If the acceleration of the elevator is $\boldsymbol{a}$ and the mass of the person standing on the scales is m, Newton's second law tells us that

$$\mathbf{F}_{net} = \mathbf{F}_s + \mathbf{F}_g = m\mathbf{a}$$

where $\mathbf{F}_g = m\mathbf{g}$ is the force of gravity acting on the person and $\mathbf{F}_s$ is the force exerted by the scales on the person. We must be careful of the vector nature of this relationship when we apply it to specific situations.

Example 9.2.1

A 90-kg person is standing on bathroom scales in an elevator that has an acceleration of 2 m/s² upward. What is the reading on the scales?

The reading on the scales is the measure of the force exerted by the scales on the person. We begin by choosing the upward direction as positive. Then a is positive if the acceleration is upward, and negative if the acceleration is downward. The sign of the acceleration due to gravity is negative, because the force of gravity is always downward. Therefore, solving our relationship for F_s, we have

$$F_s = ma - mg = m(a - g)$$
$$= (90\text{ kg})\left[2\text{ m/s}^2 - (-9.8\text{ m/s}^2)\right] = 1060\text{ N}$$

Practice: What is the scale reading if the acceleration is downward at 2 m/s?
Answer: 702 N

We can calculate the "g-forces" acting on the person by looking at the ratio of the sizes of the forces due to the scales and due to gravity.

$$g\text{-}force = \frac{F_s}{F_g}$$

As a check we know that for the special case of no acceleration, F_s and F_g have the same sizes. Therefore, the person experiences the normal force of 1 g. An alternate expression can be obtained by substituting our expressions for the two forces.

$$g\text{-}force = \frac{F_s}{F_g} = \frac{ma - mg}{-mg} = \frac{a - g}{-g}$$

Example 9.2.2

What *g-force* does the person in Example 9.2.1 experience?

The person experiences a *g-force* of

$$g\text{-}force = \frac{F_s}{F_g} = \frac{1060 \text{ N}}{882 \text{ N}} = 1.2\, g$$

or

$$g\text{-}force = \frac{a - g}{-g} = \frac{(2 \text{ m/s}^2) - (-9.8 \text{ m/s}^2)}{9.8 \text{ m/s}^2} = 1.2\, g$$

Practice: What is the *g-force* if the elevator is accelerating downward at 2 m/s^2?
Answer: 0.796 *g*

Example 9.2.3

Assume that our elevator can somehow be accelerated sideways and that the scales can be tilted so that the force of the scales acts in the proper direction to cancel out the force of gravity and provide the necessary sideways acceleration. What is the scale reading if the acceleration of the elevator is 2 m/s^2?

The vector form of our relationship is still valid.

$$\mathbf{F} = m\mathbf{a} - m\mathbf{g} = m(\mathbf{a} - \mathbf{g}) = m\mathbf{a}_{eff}$$

To find the effective acceleration, we draw the vector diagram in Figure 9.2.1. The effective acceleration $\mathbf{a}_{eff}$ is equal to the vector difference of **a** and **g**. In this case, the vectors are perpendicular to each other and the size of the vector can be found using the Pythagorean theorem.

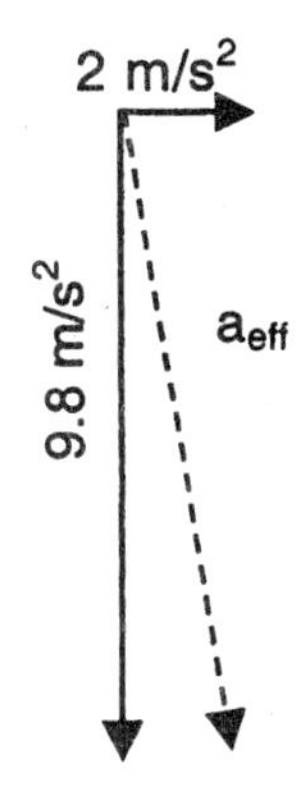

Fig. 9.2.1

$$a_{eff} = \sqrt{a^2 + g^2}$$
$$= \sqrt{\left(2\ \mathrm{m/s^2}\right)^2 + \left(9.8\ \mathrm{m/s^2}\right)^2} = 10\ \mathrm{m/s^2}$$

Therefore, the scale reading is

$$F = m a_{eff} = (90\ \mathrm{kg})\left(10\ \mathrm{m/s^2}\right) = 900\ \mathrm{N}$$

Notice that the scale reading does not increase as much as it did in Example 9.2.1. The direction of $\mathbf{F}_s$ can be measured from the diagram to be 11° from the vertical.

Practice: What is the scale reading if the sideways acceleration is increased to 5 m/s^2?
Answer: 990 N

9.3 Centrifugal Forces

If an object is at rest in a rotating reference system and we wish to explain this through the invention of a fictitious force, we conclude that the centrifugal force is equal in size to the centripetal force but oppositely directed. Therefore, we can use our equations for the size of the centripetal acceleration and centripetal force developed in Section 4.2 to calculate the corresponding sizes for the centrifugal acceleration and force.

$$a_c = \frac{v^2}{r} \qquad F_c = m\frac{v^2}{r}$$

Example 9.3.1

Assume that a space station with a radius of 1 km is rotating about its center at 1 rev/min. What is the strength of the artificial gravity produced by the centrifugal force?

We begin by calculating the speed of a point on the rim of the space station. This is just the circumference of the station divided by its rotational period, the time it takes to rotate once.

$$v = \frac{C}{T} = \frac{2\pi r}{T} = \frac{(6.28)(1000\ \mathrm{m})}{60\ \mathrm{s}} = 105\ \mathrm{m/s}$$

Then we calculate the centrifugal acceleration.

$$a_c = \frac{v^2}{r} = \frac{(105 \text{ m/s})^2}{1000 \text{ m}} = 11 \text{ m/s}^2$$

Practice: What would the acceleration be if the rotational speed is doubled?
Answer: 43.8 m/s^2. This would be a "crushing" acceleration!

9.4 Earth: A Nearly Inertial System

We do not feel Earth move and it seems that our massive planet is motionless. But, in fact, it is moving at a very high speed. Consider the annual motion of Earth around the Sun. Assuming Earth's orbit to be circular with a radius of 1.50×10^{11} m and knowing that there are about 3.16×10^7 s in 1 year, we can calculate its speed.

$$v = \frac{C}{T} = \frac{2\pi r}{T} = \frac{(6.28)(1.5 \times 10^{11} \text{ m})}{3.16 \times 10^7 \text{ s}} = 2.98 \times 10^4 \text{ m/s}$$

Although this is a fast speed, it may not seem so fast written in this form, so let's convert it to km/h.

$$2.98 \times 10^4 \frac{\text{m}}{\text{s}}\left[\frac{1 \text{ km}}{1000 \text{ m}}\right]\left[\frac{3600 \text{ s}}{1 \text{ h}}\right] = 107{,}000 \text{ km/h} \qquad (66{,}500 \text{ mph})$$

In addition, every point on Earth's surface moves in a circular path every 24 hours. A person on the equator travels about 40,000 km (24,900 miles) in 24 hours. This is a speed of 463 m/s = 1670 km/h = 1040 mph.

However, it is the acceleration, not the speed, that determines whether Earth is an inertial system. Therefore, we need to calculate the accelerations due to each of its motions to see how much each one contributes.

Let's begin with the acceleration due to Earth's rotation on its axis.

$$a_c = \frac{v^2}{r} = \frac{(463 \text{ m/s})^2}{6.37 \times 10^6 \text{ m}} = 3.37 \times 10^{-2} \text{ m/s}^2$$

Because this acceleration is only 0.34% of the acceleration due to gravity at Earth's surface, it is a relatively small effect.

The same calculation for Earth's revolution around the Sun yields

$$a_c = \frac{v^2}{r} = \frac{(2.98 \times 10^4 \text{ m/s})^2}{1.50 \times 10^{11} \text{ m}} = 5.92 \times 10^{-3} \text{ m/s}^2$$

which is smaller than the acceleration due to Earth's rotation, even though the speed of revolution is much larger than the speed for rotation.

Problems

1. A person riding on a flatcar moving at 8 m/s throws a baseball straight up at 6 m/s. What would be the baseball's initial speed relative to an observer on the ground?

2. A person riding on a flatcar throws a baseball straight up at 8 m/s. A second person observing from the ground sees the ball launched at a 45° angle with respect to the ground. How fast is the flatcar moving?

3. A spring gun fires a ball horizontally at 50 m/s. It is mounted on a flatcar moving in a straight line at 30 m/s. Relative to the ground, what is the horizontal speed of the ball when the gun is aimed to the side?

4. A train is traveling along a straight, horizontal track at a constant speed of 120 km/h. If a ball is fired directly to the side with a speed of 40 km/h relative to the train, what is its speed relative to the ground?

5. If a child weighs 250 N standing at rest on Earth, what is the weight of this child in an elevator being accelerated upward with a constant value of 2 m/s^2?

6. What is the scale reading if a 30-kg dog lies on the scale in an elevator accelerating upward at 2 m/s^2?

7. What is the weight of the child in Problem 5 if the acceleration is downward?

8. What is the scale reading for the dog in Problem 6 if the acceleration is downward?

9. An observer in a train with an acceleration of 3 m/s^2 notices that a ball falls in a straight line that is slanted toward the back of the train. What is the acceleration of the ball along this line?

10. What is the angle of the ball's path in the previous problem?

11. If a passenger on a jet that is accelerating down the runway drops a set of keys from rest, they will travel in a straight line directed down and back. Draw a figure depicting the path of the keys if instead they are initially thrown (a) straight down, or (b) toward the rear of the jet.

12. You are in a jet that has just landed and is rapidly slowing down. You reach out into the isle and throw your keys toward a spot on the floor that is 45° down and forward of your hand. The keys appear to arc slightly upward and land beyond the spot in the floor. Is the magnitude of the plane's acceleration less than, equal to, or greater than *g*?

*13. A ball is dropped from a height of 2 m in a train traveling at 30 m/s and accelerating at 2 m/s^2. How far will the ball miss the spot directly below where it was dropped?

*14. A ball is dropped from a height of 1 m in a train traveling at 20 m/s and accelerating at 1 m/s^2. How far will the ball miss the spot directly below where it was dropped?

15. What is the acceleration at the rim of a merry-go-round if it has a radius of 3 m and makes one revolution every 4 s? How does this value compare to the acceleration due to gravity?

16. What is the acceleration due to artificial gravity if a space station has a radius of 1000 m and rotates at 0.5 rpm? How does this compare to the value of *g* on Earth?

17. A 60-kg person is riding on the rotating cylinder ride shown in Figure 9-9. If the person's speed is 7 m/s and he feels a 1500-N centrifugal force, what is the radius of the cylinder?

18. A 600-N person is riding on the rotating cylinder ride shown in Figure 9-9. With the ride in operation, the magnitude of her weight increases to 1000 N. If the radius of the cylinder is 2 m what is her speed?

19. What is the centrifugal acceleration on the equator of the Moon if it has a rotational period of 27.3 days?

20. What is the centrifugal acceleration on the equator of Venus if it has a rotational period of 243 days?

21. Calculate Earth's acceleration due to its motion around the center of the Milky Way Galaxy. Assume that the orbital radius is about 3×10^{17} km and the orbital period is 2.5×10^{8} years. How does this acceleration compare to the value of g?

22. What is the acceleration of Mars around the Sun?

10 — EINSTEIN'S RELATIVITY

10.1 Searching for the Medium of Light

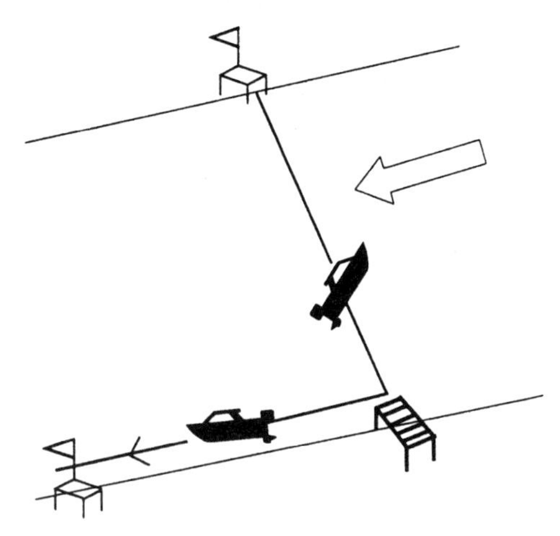
Fig. 10.1.1

The predicted results for the Michelson-Morley experiment can be illustrated using a more common example of two boats racing along perpendicular paths on a river. Here, the analog of the ether wind is the river's current. The boats start at the same point and travel equal distances at equal speeds relative to the water but in different directions. We will use the symbol c (normally reserved for the speed of light) for the boats' speeds and v for the speed of the current to further highlight the analogy.

Suppose the first boat's path is parallel to the current, and the second boat's path is perpendicular to the current. The first boat travels downstream a distance L, turns around instantaneously, and returns to the starting point, as shown in Figure 10.1.1. The second boat travels straight across the river, again a distance L, and returns.

For the first boat the time t_{1d} to travel downstream is obtained by dividing the distance traveled by the speed of the boat relative to the shore (see Section 9.1). This speed is the sum of the speed of the boat relative to the water plus the speed of the water relative to the shore. If we assume a boat speed of 3 m/s, a current of 1 m/s, and a distance of 1200 m, we obtain

$$t_{1d} = \frac{L}{c+v} = \frac{1200\text{ m}}{3\text{ m/s} + 1\text{ m/s}} = 300\text{ s}$$

The time t_{1u} that it takes the boat to return upstream is given by

$$t_{1u} = \frac{L}{c-v} = \frac{1200\text{ m}}{3\text{ m/s} - 1\text{ m/s}} = 600\text{ s}$$

This gives a total time t_1 for the round trip of 900 s.

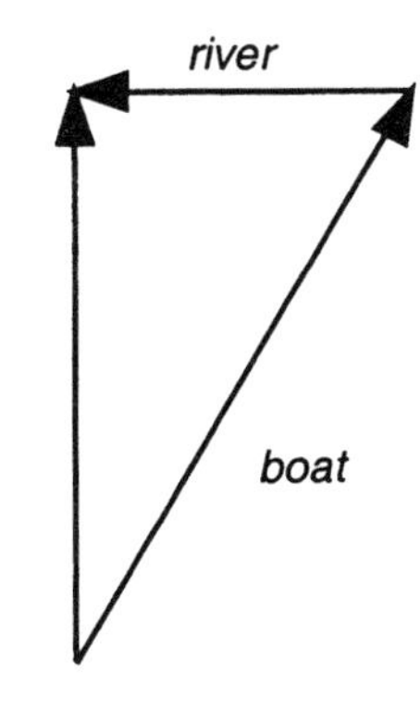

Fig 10.1.2

In order for the second boat to travel straight across the river, it must aim upstream. This reduces its speed across the river. We can find this speed by requiring that the velocity of the boat plus the velocity of the current add to point straight across the river, as shown in Figure 10.1.2. The speed of 2.83 m/s can be obtained graphically as we did in Chapter 4 in the text or by using the Pythagorean theorem (Section 4.1). Using this value, we calculate the time t_2 to travel across the river to be

$$t_2 = \frac{1200\text{ m}}{2.83\text{ m/s}} = 424\text{ s}$$

Because the trip back takes the same amount of time, the total round-trip time is 848 s. Therefore, the second boat arrives back at the finish line 52 s before the first boat, even though the boats are identical.

The puzzle for physicists at the turn of the century was that the light beams came back at the same time!

Fig. 10.2.1

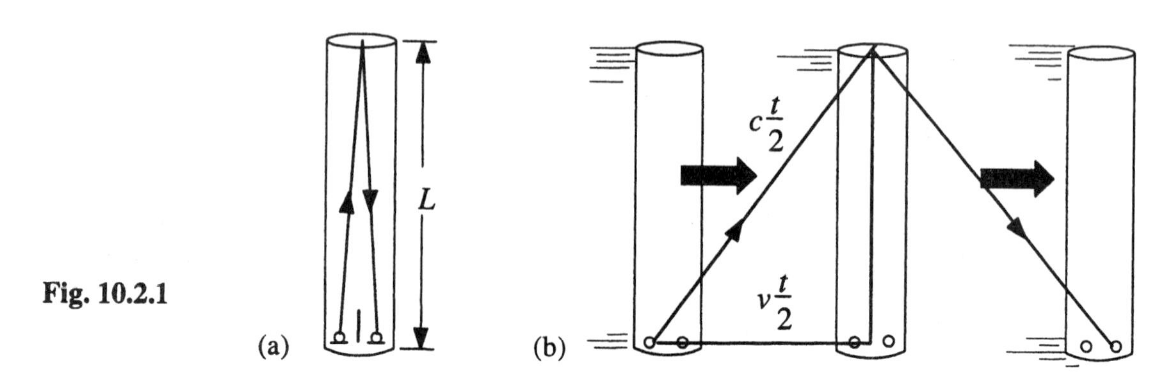

10.2 Experimental Evidence for Time Dilation

The size of the time-dilation effect can be calculated precisely using the Pythagorean theorem. First, let's consider the time as seen by an observer at rest relative to the light clock we used in the text. In this reference system the light beam travels straight up to the mirror and returns as shown in Figure 10.2.1(a). It travels a distance equal to twice the length of the light clock. This requires a time t_o

$$t_o = \frac{2L}{c}$$

The distance traveled by the light beam in the moving reference system depends on the relative speed of the moving system. Figure 10.2.1(b) shows the distances the light beam travels (as seen by an observer in the rest system) in terms of the speed multiplied by the total time t. The Pythagorean theorem for the left-hand triangle is

$$\left[\frac{ct}{2}\right]^2 = \left[\frac{vt}{2}\right]^2 + L^2$$

To obtain a relationship between the times in the two systems, we substitute for L from the at-rest system to obtain

$$\left[\frac{ct}{2}\right]^2 = \left[\frac{vt}{2}\right]^2 + \left[\frac{ct_o}{2}\right]^2$$

Solving for the time t in the moving system yields the amount that the time is dilated as viewed from the rest system.

$$t = \frac{1}{\sqrt{1-(v/c)^2}}\,t_o = \gamma t_o$$

where γ is the adjustment factor for special relativity. Some values of γ are listed in Table 10-1 in the text.

Example 10.2.1

How much is the time dilation for $v = \frac{1}{2}c$?

Let's begin by calculating the adjustment factor γ.

$$\gamma = \frac{1}{\sqrt{1-(v/c)^2}} = \frac{1}{\sqrt{1-(\frac{1}{2})^2}} = \frac{1}{\sqrt{\frac{3}{4}}} = \sqrt{\tfrac{4}{3}} = 1.15$$

Therefore $t = 1.15t_0$. This means that an event that takes 1 s in the at-rest system is viewed as taking 1.15 s when viewed from the moving system. It is important to realize that this works both ways. One second for observers in the moving system lasts 1.15 s when viewed by observers in the at-rest system.

Practice: What is the adjustment factor for a speed of 0.75 c?
Answer: 1.51

Example 10.2.2

What speed is required for time to be slowed by a factor of 2?

We start with our expression for the adjustment factor, solve it for v, and set $\gamma = 2$.

$$v = c\sqrt{1-\frac{1}{\gamma^2}} = c\sqrt{1-\frac{1}{2^2}} = c\sqrt{\tfrac{3}{4}} = 0.866\,c$$

Practice: What speed is required for a time dilation by a factor of 4?
Answer: 0.968 c

10.3 Comparing Velocities

We have already seen that the second postulate requires a new rule for connecting velocities measured in two different inertial reference systems. Although the derivation of the new rule is beyond the mathematical level of this textbook, we give the rule to illustrate how two velocities close to the speed of light can be added without exceeding the speed of light.

Assume that the relative speed of system #2 as measured in system #1 is v as shown in Figure 10.3.1. Assume further that our object has speed u_2 as measured in system #2 and that the object is moving parallel to the relative velocity. Then the speed u_1 measured in the first system is

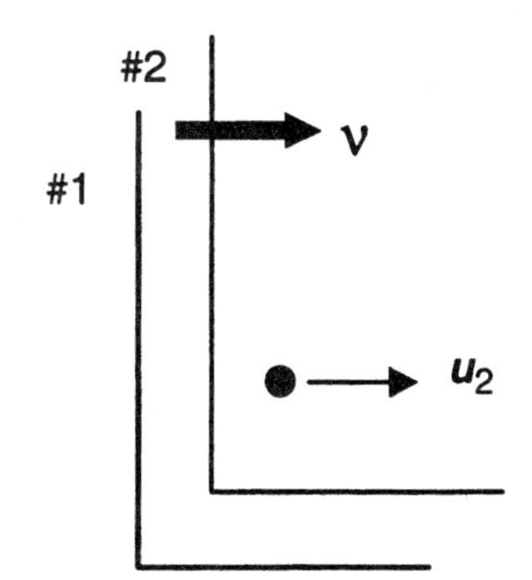

Fig. 10.3.1

$$u_1 = \frac{u_2 + v}{1 + \frac{u_2 v}{c^2}}$$

This rule has the same form as the classical rule except for the term in the denominator. Notice that the rule reduces to the classical rule if u_2 is very much less than c because the value of the denominator is very close to 1. This must be the case, since the classical rule worked very well for small velocities.

Example 10.3.1

It is interesting to look at the extreme situation where the speed in the moving system is c and the system is moving with any speed v. What is u_1?

We set $u_2 = c$ in the rule.

$$u_1 = \frac{u_2 + v}{1 + \frac{u_2 v}{c^2}} = \frac{c + v}{1 + \frac{cv}{c^2}} = \frac{c+v}{\frac{1}{c}(c+v)} = c$$

Thus, we find that u_1 equals c as required by the second postulate.

Practice: If the object has a speed of $\frac{1}{2}c$ in system #2 and the relative speed is $\frac{1}{2}c$, what is the speed of the object in system #1?

Answer: 0.8 c

10.4 Length Contraction

The length of a moving object L measured along its direction of motion is equal to its length L_o measured by an observer at rest relative to the object divided by the relativistic adjustment factor. Therefore,

$$L = \frac{L_o}{\gamma}$$

Example 10.4.1

How long is a meter stick moving at 0.8 c along its length?

$$L = \frac{L_o}{\gamma} = L_o\sqrt{1 - (v/c)^2} = (1\,\text{m})\sqrt{1 - 0.8^2} = 0.6\,\text{m}$$

Practice: How long will the meter stick be if its speed is 0.9 c?

Answer: 0.436 m

10.5 Relativistic Laws of Motion

The classical form of Newton's second law carries over into special relativity if we write it in terms of momentum and use the relativistic form for the momentum.

$$F = \frac{\Delta p}{\Delta t} = \frac{\Delta(\gamma m v)}{\Delta t}$$

Example 10.5.1

A proton in a moderately strong electric field experiences an electric force of 10^{-14} N. How long would it take this field to accelerate the proton from rest to 80% of the speed of light?

At this speed the relativistic adjustment factor is equal to 1.67. We solve the relativistic form of Newton's second law for the time *t*, set the initial momentum equal to zero, and plug in the known values.

$$t = \frac{\gamma m v}{F} = \frac{(1.67)(1.67 \times 10^{-27}\ \text{kg})(0.8 \times 3 \times 10^{8}\ \text{m/s})}{10^{-14}\ \text{N}} = 66.9\ \mu\text{s}$$

If we increase the final speed to 90% of *c*, the time increases to 103 μs. Therefore, it requires 36 μs to increase the speed from 80% to 90% *c*. It takes half as long to gain the last 10% as it did to gain the original 80%. If we continue this trend, it will require an additional 50 μs to increase the speed from 90% to 95% *c*, half the gain in a longer time. Each additional gain requires longer and longer to accomplish, increasing very rapidly as we approach *c*, the limiting speed.

Practice: How long would it take this force to accelerate the proton from rest to 0.99 *c*?
Answer: 352 μs

Example 10.5.2

What is the rest-mass energy of a proton?

$$E_o = mc^2 = (1.67 \times 10^{-27}\ \text{kg})(3 \times 10^{8}\ \text{m/s})^2 = 1.5 \times 10^{-10}\ \text{J}$$

Example 10.5.3

What is the total energy of a proton traveling at 0.9 *c*?

$$E = \gamma m c^2 = (2.29)(1.67 \times 10^{-27}\ \text{kg})(3 \times 10^{8}\ \text{m/s})^2 = 3.44 \times 10^{-10}\ \text{J}$$

Note that this answer is larger than that in Example 10.5.2 by the relativistic adjustment factor. If you use the concept of a relativistic mass, the proton traveling at 0.9 *c* has a mass that is 2.29 times larger that when it is a rest.

Practice: What is the total energy of a proton traveling at 0.95 *c*?
Answer: 4.81×10^{-10} J

10.6 Gravity and Acceleration

The equivalence principle states that *a constant acceleration is completely equivalent to a uniform gravitational field.* This means that doing physics in a spaceship with a constant acceleration in deep space is just like doing physics on the surface of a planet with an acceleration due to gravity equal to the acceleration of the spaceship. Therefore, being in a spaceship with an acceleration of 9.8 m/s^2 is just like being on Earth. The direction of this relativistic gravitational field is always opposite the direction of the acceleration. If the acceleration takes place in the presence of a gravitational field, we can find the effective gravity g_{eff} by combining the actual gravitational field g with that due to the acceleration g_{rel}.

Example 10.6.1

A train is traveling down a long, straight track with a constant acceleration of 5 m/s^2 in the forward direction. If a passenger in the train drops a ball, what is the path of the ball and how large is its acceleration as viewed in the train?

Fig. 10.6.1

The acceleration of the train is equivalent to a gravitational field g_{rel} in the backward direction that produces an acceleration of the ball equal to 5 m/s^2. We must add this to the normal gravitational field on Earth that produces an acceleration of 9.8 m/s^2 downward.

Remembering that both of these are vectors, we can add them using our graphical method as shown in Figure 10.6.1. The path is obtained by measuring the angle on the graph. The ball falls along a straight line tilted backward 27° from the vertical.

We can also obtain the magnitude of the effective acceleration using the Pythagorean theorem.

$$g_{eff} = \sqrt{g_{rel}^2 + g^2} = \sqrt{\left(5\text{ m/s}^2\right)^2 + \left(9.8\text{ m/s}^2\right)^2} = 11\text{ m/s}^2$$

Practice: At what angle and with what acceleration will the ball fall if the train's acceleration is 9.8 m/s^2?
Answer: 13.9 m/s^2 at 45° from the vertical

10.7 Gravity and Light

Example 10.7.1

How much does a horizontal beam of light fall in 9 m?

As this is a problem in projectile motion, we begin by calculating the time for the light to travel 9 m horizontally.

$$t = \frac{d}{v} = \frac{9\text{ m}}{3 \times 10^{8}\text{ m/s}} = 3 \times 10^{-8}\text{ s}$$

We can now plug this time into our equation for the distance fallen from rest in free fall.

$$d = \tfrac{1}{2}gt^{2} = \tfrac{1}{2}\left(9.8\text{ m/s}^{2}\right)\left(3 \times 10^{-8}\text{ s}\right)^{2} = 4.41 \times 10^{-15}\text{ m}$$

Because the diameter of an atom is typically 10^{-10} m, the light beam falls much less than the diameter of an atom, as stated in the text.

Problems

1. Cindy can paddle a canoe at a speed of 2 m/s in flat water. She wishes to cross a 100-m wide river with a current of 0.5 m/s. She aims the canoe straight across the river. How long does it take her to cross the river and how far downstream does she land?

2. Cindy from the previous problem decides instead to aim the canoe upstream so that she lands directly across the river from her starting point. How long does it take her to cross the river?

3. A boat race takes place on a river with a current of 5 km/h. If the boats have speeds of 30 km/h relative to the water, how long would it take to complete a race that goes 50 km up the river and back? How much longer does it take than for the same race on a lake?

*4. Knowing that the orbital speed of Earth is 100,000 km/h, what is the maximum time difference Michelson and Morley could have expected if each arm of their apparatus were 1 m long?

5. By what factor is time dilated for a speed of 0.95 *c*?

6. How much time would a clock lose in 1 year if it were traveling at 0.2 *c*?

*7. Imagine an astronaut orbiting Earth in the International Space Station at an orbital speed of 28,000 km/h. If the astronaut remains in orbit for one year as measured by a clock on Earth, how much younger would the astronaut be than if she stayed home. The adjustment factor for this speed is about 1.000 000 000 34. [*Hint*: To simplify the algebra, it will be helpful to use the approximation $1/(1+\beta) \approx (1-\beta)$, when β is much less than one.]

*8. The average orbital speed of Earth as it revolves around the Sun is about 100,000 km/h. In an ill-informed attempt to find a "fountain of youth," Skip Parsec decides to spend a year hovering at a fixed point in space and wait for Earth to return to that point. Will Skip be younger or older than if he had stayed home? By how much? The adjustment factor for this speed is about 1.000 000 004 3.

9. If you wanted to travel a distance of 10 light-years while aging only 1 year, how fast would you need to travel?

10. The idea behind the movie *Buck Rogers in the 25th Century* is theoretically possible. With what speed would Buck have to travel in order to age only 5 years in 500 years of traveling?

11. If a particle has a speed of 0.5 c toward the front of a spaceship that is traveling away from Earth at 0.8 c, what is the speed of the particle relative to Earth?

12. What is the speed of the particle in the previous problem if the particle is traveling toward the back of the spaceship?

13. A particle traveling at 0.98 c in the laboratory decays by emitting a particle with a speed of 0.4 c measured in the particle's reference system. If the decay is in the forward direction, what is the speed of the decay particle in the laboratory?

14. What is the speed of the decay particle in the previous problem if the decay is in the backward direction?

15. How far away is Uranus when observed from a spaceship near Earth traveling at 90% of the speed of light? The distance measured from Earth at this time is 2.8×10^{12} m.

16. What is the length of a meter stick traveling at 0.8 c in a direction along its length?

*17. How fast would a meter stick have to be moving to be half its original length?

*18. If you want to fit a 10-m pole into a 4-m shed, how fast must it be moving?

19. What impulse is needed to accelerate (a) a proton, and (b) an electron from rest to 0.9 c?

20. What impulse is needed to accelerate a proton from moving at 0.9 c to the right to moving at 0.9 c to the left?

21. What average force is needed to accelerate a proton from rest to 0.9 c in 1 μs?

22. What average force is needed to accelerate an electron from rest to 0.9 c in 1 μs?

23. What is the rest-mass energy of a neutron?

24. What is the rest-mass energy of an electron?

25. If the total energy of a muon is 4 times its rest-mass energy, how fast must it be moving?

26. How fast must a proton be moving for its rest-mass energy to be only 10% of its total energy?

*27. By what factor does the relativistic momentum increase when the speed doubles from 0.3 c to 0.6 c?

*28. By what factor does the relativistic momentum increase when the speed doubles from 0.4 c to 0.8 c?

29. A fast-growing bean stock grows in a train with a constant acceleration of 2 m/s^2. What angle does the stock make with the floor?

30. What is the effective gravity in a train with a constant acceleration of 3 m/s^2?

31. By how much would a horizontal beam of light fall while traveling across a 20-m wide banquet room?

32. If the speed of light were only 4000 m/s, by how much would a horizontal beam of light fall while traveling across a 20-m wide room?

33. When our Sun eventually can no longer sustain nuclear fusion in its core, it will throw off its outer layers leaving behind a stellar remnant known as a white dwarf. White dwarfs are typically the size of our Earth while containing about the mass of our Sun. What is the acceleration due to gravity at the surface of a typical white dwarf?

34. What is the acceleration due to gravity at the surface of a neutron star that has a mass of 4×10^{30} kg (twice that of the Sun) and a radius of 10 km?

11 — STRUCTURE OF MATTER

11.1 Masses and Sizes of Atoms

Joseph Gay-Lussac reported in 1809 "Whenever gases react or gases form under the same conditions of temperature and pressure they do so in the ratio of small whole numbers." The example given in the text illustrates this: One liter of nitrogen (N_2) combines with 3 L of hydrogen (H_2) to form 2 L of ammonia (NH_3). John Dalton had a very difficult time accepting these results. He was convinced that atoms were the basic unit and that atoms of the same kind could not react with each other. If equal volumes of gas contained equal numbers of atoms, then only one liter of ammonia should be produced. In 1811 Amedao Avogadro published what is now known as Avogadro's hypothesis, which said, "Equal volumes of gases under the same conditions of temperature and pressure contain equal numbers of molecules". The results of Gay-Lussac were consistent with this hypothesis, providing hydrogen gas and nitrogen gas were each made up of diatomic molecules. Later evidence confirmed that this is so.

Example 11.1.1

Carbon monoxide (CO) released during the combustion of gasoline reacts with oxygen (O_2) in the hot catalytic converter to form carbon dioxide (CO_2). What volume of CO_2 is formed from the oxidation of 14 L of CO?

In the oxidation process each CO molecule acquires another oxygen atom. This requires that two CO molecules react with every one molecule of O_2. This reaction produces two molecules of CO_2.

$$2CO + O_2 \Rightarrow 2CO_2$$

Therefore, 14 L of carbon monoxide will combine with 7 L of oxygen to produce 14 L of carbon dioxide.

Practice: Ozone (O_3) can be broken down to form oxygen. How many liters of oxygen will result from the breakdown of 24 L of ozone?
Answer: 36

Oil floats on water and, given enough surface, can spread out making a very thin film. Oleic acid, a compound of hydrogen, carbon, and oxygen, does not mix with water and spreads out even better than oil. A single drop of oleic acid will cover the surface of an entire swimming pool. If we mix this compound with alcohol and put a single drop of the alcohol/acid mixture on water, we have a way of putting a very, very small amount of oleic acid on the water's surface. The alcohol evaporates very fast and only the oleic acid is left on the water's surface. If we first sprinkle fine flour on the water's surface, we see that the acid pushes the flour away as it spreads out into a circular shape that we can measure.

Suppose we mix 1 part of oleic acid with 499 parts of alcohol. We then measure the volume of a single drop of this solution by putting a number of drops into a graduated beaker. The volume of the oleic acid by itself V_{acid} is the volume of the drop V_{drop} divided by 500.

If the experiment gives a circular pattern with a radius r, we can calculate the circle's area A using the relationship $A = \pi r^2$. This area is only due to the acid layer since the alcohol evaporated. Multiplying the area by the height h of the thin film also gives the volume of the oleic acid. Setting the two volumes of oleic acid equal to each other gives us a way to obtain the height.

$$V_{acid} = \frac{V_{drop}}{500} = Ah$$

The only unknown is the height h, so we can solve the equation for the height.

$$h = \frac{V_{acid}}{A} = \frac{V_{drop}}{500\,A}$$

This gives us an upper limit on the size of an oleic acid molecule, as the layer must be at least one molecule thick.

Example 11.1.2

If it takes 135 drops from our eye dropper to yield a volume of 6 cm^3, and if an experiment gives a circular pattern with a radius of 15 cm, what is the thickness of the layer?

A single drop of the alcohol/acid mixture has a volume of

$$V_{acid} = \frac{V_{drop}}{500} = \frac{6\text{ cm}^3}{135 \times 500} = 8.89 \times 10^{-5}\text{ cm}^3$$

The area of the circle is

$$A = \pi r^2 = 3.14\left(15\text{ cm}\right)^2 = 707\text{ cm}^2$$

We can obtain the thickness of the layer by dividing the volume by the area.

$$h = \frac{V_{acid}}{A} = \frac{8.89 \times 10^{-5}\text{ cm}^3}{707\text{ cm}^2} = 1.26 \times 10^{-7}\text{ cm}$$

Practice: If the circular pattern was incorrectly measured and the 15 cm is actually the diameter of the circle, what is the height of the molecule?
Answer: Decreasing the radius of the circle reduces the area by a factor of four, making our estimate of the molecule's height four times bigger.

If we assume that the oleic acid molecule is roughly cubical, this height is also the width and length of the molecule. Cubing this number gives an estimate of the volume V of a single oleic acid molecule.

$$V_{molecule} = lwh = \left(1.26 \times 10^{-7}\ \text{cm}\right)^3 = 2 \times 10^{-21}\ \text{cm}^3$$

Dividing the molecular volume into the macroscopic volume gives the number of molecules N in our sample.

$$N = \frac{V_{acid}}{V_{molecule}} = \frac{8.89 \times 10^{-5}\ \text{cm}^3}{2 \times 10^{-21}\ \text{cm}^3/\text{molecule}} = 4.44 \times 10^{16}\ \text{molecules}$$

We can estimate the mass of the oleic acid molecule from its density. The density of the oleic acid is about 0.9 g/cm^3. Because density is an intrinsic property of matter, this must be approximately the density of a single molecule.

$$m = DV = \left(0.9\ \text{g/cm}^3\right)\left(2 \times 10^{-21}\ \text{cm}^3\right) = 1.8 \times 10^{-24}\ \text{kg}$$

How good are our assumptions? First, we assumed that the molecules would form a layer one molecule thick. If they didn't, our value for h is only an upper limit on the size of the oleic acid molecule. The assumption that the molecules are cubical in shape is known to be incorrect. The molecule is a long chain that is about 100 atoms long. Furthermore, they tend to line up perpendicular to the surface. This makes the diameter of the chain less than 1 nm.

Obviously, the mass and size of a single hydrogen, carbon, or oxygen atom is much smaller, but this experiment gives a simple, though indirect, way of estimating molecular sizes and masses.

11.2 Pressure

The text gives the pressure as a macroscopic property and describes the microscopic origins of pressure in a gas. Using the concepts of impulse and momentum changes developed in Chapter 6 and the concept of kinetic energy from Chapter 7, we can get an exact expression that connects the macroscopic, observable pressure with its microscopic origins, namely atomic motion. The derivation is a little messy but has only a few assumptions, and is, thus, relatively straightforward. The derivation can be found in all introductory engineering physics textbooks.

$$P = \frac{2}{3}\frac{N}{V}\left(\tfrac{1}{2}m\overline{v^2}\right)$$

This says that the pressure is directly proportional to the number of particles N and the average kinetic energy of the particles. It is inversely proportional to the volume V of the container holding the gas.

Example 11.2.1

What is the pressure created in a 1-m^3 box by 1 million molecules of oxygen with an average speed of 500 m/s?

Because the oxygen molecule has two atoms, the mass of the oxygen molecule is 32 amu. The kinetic energy of each oxygen molecule is

$$KE = \tfrac{1}{2}mv^2 = \tfrac{1}{2}\left(32 \times 1.66 \times 10^{-27}\ \text{kg}\right)\left(500\ \text{m/s}\right)^2 = 6.64 \times 10^{-21}\ \text{J}$$

The pressure then becomes

$$P = \frac{2}{3}\frac{N}{V}\left(\tfrac{1}{2}mv^2\right) = \frac{2}{3}\frac{10^6}{1\,\text{m}^3}\left(6.64 \times 10^{-21}\ \text{J}\right) = 4.43 \times 10^{-15}\ \frac{\text{N}}{\text{m}^2}$$

This is a very small pressure, only 4.37×10^{-20} atm, because the number of molecules is very small. A box this size would contain about 3×10^{25} molecules at atmospheric pressure.

Practice: What happens to the pressure of a gas when the speed of the molecules is doubled?
Answer: Doubling the speed quadruples the kinetic energy and, therefore, quadruples the pressure.

11.3 Temperature

To convert from a Fahrenheit temperature to a Celsius temperature, we have to subtract 32 degrees from the Fahrenheit reading to get to the zero point on the Celsius scale and then adjust for the different size degrees. Because there are 180 Fahrenheit degrees and 100 Celsius degrees between the freezing point and the boiling point of water, we can calculate the relative sizes of the two degrees.

$$\frac{100°\text{C}}{180°\text{F}} = \frac{5°\text{C}}{9°\text{F}}$$

Therefore, the equation for converting °F to °C is

$$T_C = \frac{5°\text{C}}{9°\text{F}}\left(T_F - 32°\text{F}\right)$$

where T_C and T_F are the corresponding temperatures in Celsius and Fahrenheit. When writing this equation, it helps to remember that the conversion factor has to have °F in the denominator to cancel the °F in the parentheses. To decide whether to subtract or add the 32°F, remember that the freezing point of water is 0°C = 32°F.

Converting from °C to °F is the reverse of this.

$$T_F = \frac{9°\text{F}}{5°\text{C}}T_C + 32°\text{F}$$

Example 11.3.1

What is the temperature in Celsius when your Fahrenheit thermometer reads 100°?

$$T_C = \frac{5°\text{C}}{9°\text{F}}\left(T_F - 32°\text{F}\right) = \frac{5°\text{C}}{9}\left(100 - 32\right) = 37.8°\text{C}$$

Practice: What is the Celsius temperature if your thermometer reads one-half this value, or 50°F?
Answer: 10°C

Example 11.3.2

A European friend writes to you that last summer the temperature in his city reached a high of 40°C. What is this temperature in Fahrenheit?

$$T_F = \frac{9°\text{F}}{5°\text{C}} T_C + 32°\text{F} = \frac{9°\text{F}}{5} 40 + 32°\text{F} = 104°\text{F}$$

Practice: Room temperature is 20°C. What is it in Fahrenheit?
Answer: 68°F

The conversion from the Celsius temperature scale to the absolute temperature scale is easy because the degrees are the same size, that is, 1°C = 1 K. You just need to add or subtract 273°. If you are converting a Celsius reading to an absolute (or Kelvin) reading, you add the 273°. To convert from Fahrenheit, you must first convert to Celsius.

Example 11.3.3

Oxygen boils at 90 K. What is this temperature in Celsius?

$$T_C = T_K - 273°\text{C} = 90\text{ K} - 273°\text{C} = -183°\text{C}$$

Practice: Room temperature is about 20°C. What is this temperature on the absolute scale?
Answer: 293 K

11.4 The Ideal Gas Law: A Microscopic View

In Section 11.2, we connected the pressure as a macroscopic property to the average kinetic energy of the microscopic particles. We can go one step further and look at the microscopic origins of temperature. If we multiply both sides of the pressure relationship by the volume of the gas, we get

$$PV = \tfrac{2}{3} N \left(\tfrac{1}{2} m\overline{v^2}\right)$$

The text states that the ideal gas law is given by

$$PV = nRT$$

where the temperature T must be in kelvin.

Comparing these two equations shows that the macroscopic property we call temperature is directly proportional to the average kinetic energy of the atomic or molecular particles.

$$T \propto \tfrac{1}{2} m \overline{v^2}$$

Therefore, if two gases are at the same temperature, their molecules have equal average kinetic energies.

$$\tfrac{1}{2} m_1 \overline{v_1^2} = \tfrac{1}{2} m_2 \overline{v_2^2}$$

This allows us to calculate the ratio of their average speeds. (What we actually calculate is a special speed known as the *root mean square* (rms) speed. We calculate the square of the speed for each particle, take the average, and then take the square root.)

$$\frac{v_1}{v_2} = \sqrt{\frac{m_2}{m_1}}$$

Example 11.4.1

How does the rms speed of hydrogen molecules compare to the rms speed of oxygen molecules in a mixture of the two gases?

In a mixture the two gases are at the same temperature. Thus, we can use our relationship.

$$\frac{v_h}{v_o} = \sqrt{\frac{m_o}{m_h}} = \sqrt{\frac{32 \text{ amu}}{2 \text{ amu}}} = 4$$

Therefore, the hydrogen molecules have 4 times the rms speed of the oxygen molecules.

11.5 The Ideal Gas Law: A Macroscopic View

The ideal gas law can be rewritten to show that if we keep the amount and type of gas fixed, the ratio PV/T has a constant value.

$$\frac{PV}{T} = nR$$

This means that the three macroscopic quantities—pressure, volume, and temperature—are not independent quantities. We can obtain any one of them if we know the values of the other two.

Because PV/T is a constant and using the subscripts i and f for the initial and final values, we must have

$$\frac{P_f V_f}{T_f} = \frac{P_i V_i}{T_i}$$

Whenever any one of the quantities is held fixed, it can be canceled and we get a relationship between the other two. This gives us the various gas laws described in the text.

Example 11.5.1

What happens to the volume of 50 cm^3 of gas if its temperature is raised from 20°C to 100°C? Assume that the pressure remains the same.

Canceling the pressure, solving for the final volume, and expressing the temperatures in kelvin, we have

$$V_f = V_i\left(\frac{T_f}{T_i}\right) = \left(50\ \text{cm}^3\right)\left(\frac{373\ \text{K}}{293\ \text{K}}\right) = 63.7\ \text{cm}^3$$

Notice that any units can be used for the volume. The answer will have the same units as that of the initial volume. On the other hand, the temperature must always be expressed on the absolute scale.

Practice: What is the volume if the temperature is raised another 80°C?
Answer: 77.3 cm^3

Example 11.5.2

Suppose we have a gas in a cylinder with a movable piston. We measure its pressure, volume, and temperature at the beginning of an experiment. We heat the gas to a new temperature that is three times the original temperature (measured in kelvin). If the volume increases to twice the original volume, what is the new pressure?

We solve for the final pressure and plug in $T_f = 3T_i$ and $V_f = 2V_i$ to obtain

$$P_f = P_i\left(\frac{V_i}{V_f}\right)\left(\frac{T_f}{T_i}\right) = P_i\left(\frac{V_i}{2V_i}\right)\left(\frac{3T_i}{T_i}\right) = \frac{3}{2}P_i$$

Practice: What would the new pressure be in the above example if the volume is held fixed during the heating process?
Answer: The final pressure will be 3 times the initial pressure.

Problems

1. What volume of nitrogen is required to produce 80 liters of ammonia?

2. What volume of carbon dioxide is produced when 10 L of carbon monoxide combines with oxygen?

3. Methane gas (CH_4) can be broken down into solid carbon and hydrogen gas (H_2). What volume of hydrogen can be obtained from 100 L of methane?

4. The acetylene gas used by welders is oxidized through the reaction:

$$2C_2H_2 + 5O_2 \Rightarrow 4CO_2 + 2H_2O$$

Suppose the tanks of acetylene and the tanks of oxygen have the same pressure, volume, and temperature. How many tanks of oxygen would be needed to burn one tank of acetylene?

5. An artist uses 1000 cm^3 of paint (about a quart) to uniformly cover a canvas with an area of 10 m^2. How thick is the average coating of the paint?

6. If a painter uses 40 L (1000 cm^3 each) to paint a house with a surface area of 300 m^2, what is the thickness of the paint?

7. One liter (1000 cm^3) of oil is dumped into a gulf. Assuming that this will spread out evenly over the water to a thickness of 10^{-4} cm, how big an area will the oil cover? How does this compare to the size of a football field?

8. How much paint is required to cover a 400-m^2 barn to a thickness of 0.01 mm?

9. If an atom has a diameter of 0.2 nm, how many of them would it take to form a layer one atom thick on a plate that has an area of 50 cm^2?

10. A cube of gold 1 cm on a side has a mass of 19.3 g and contains 5.9×10^{22} atoms. What is the mass of each gold atom?

11. If 5×10^{22} nitrogen molecules with an rms speed of 450 m/s occupy a volume of 2×10^3 cm^3, what is the pressure? Nitrogen molecules have a mass of 28 amu.

12. Carbon dioxide has a mass of 44 amu. If 10^{22} carbon dioxide molecules have an average speed of 400 m/s and occupy a container with a volume of 500 cm^3, what is the pressure of the gas?

*13. Typically, the absolute pressure in a car tire is about 3 atm and at room temperature the gas in the tire will have about 8×10^{25} particles per cubic meter. Given that air is mostly nitrogen, what is the average velocity of the nitrogen molecules in the tire? How would this compare to the average velocity of the nitrogen molecules on the outside of the tire?

*14. Hydrogen molecules have a mass of 2 amu. If the average speed of the hydrogen molecules is 1600 m/s, how many hydrogen molecules per liter would you need to have a pressure of 1 atm?

15. Convert each of these Fahrenheit temperatures to the Celsius scale: −40°F, 0°F, and 70°F.

16. The record high temperature on Earth is 136°F. What is this temperature on the Celsius scale?

17. Convert each of these Celsius temperatures to the Fahrenheit scale: −40°C, 20°C, and 37°C.

18. The record low temperature on Earth is −88°C. What is this temperature on the Fahrenheit scale?

*19. A student decides to devise a new temperature scale with the freezing and boiling points of water at 0°X and 50°X. What Celsius and Fahrenheit temperatures correspond to a temperature of 30°X?

*20. What is the value of absolute zero on the temperature scale introduced in the previous problem?

*21. The freezing and boiling points of water on the Rankine temperature scale are 492°R and 672°R, respectively. What is normal body temperature on this scale?

22. If you were "running a four degree fever" on the Fahrenheit scale, what fever would you

be "running" on the Rankine scale introduced in the previous problem?

23. Convert each of the following temperatures to the Celsius scale: 0 K and 310 K.

24. The melting point of aluminum is 933 K. What is aluminum's melting point on the Celsius and Fahrenheit scales?

25. The mass of a nitrogen molecule is 0.88 times that of an oxygen molecule. If the average speed of a nitrogen molecule is 511 m/s at 20°C, what is the average speed of an oxygen molecule at this temperature?

26. The mass of an carbon molecule is 12 times that of a hydrogen molecule. How do the average speeds of these molecules compare if the gases are at the same temperature?

27. Two liters of gas at 20°C are heated to 100°C while the pressure is maintained at 1 atm. What is the final volume of the gas?

28. An ideal gas is initially at 20°C. To what temperature would you have to change the gas to double its volume at constant pressure?

29. If you hold the temperature of an ideal gas constant, what happens to its volume when you triple its pressure?

30. An ideal gas at a pressure of 2 atm is contained in a cylinder, which is immersed in a bath of ice water. A piston on the end of the cylinder is compressed slowly enough that the gas remains at the same temperature as the ice water. If the volume is reduced to one quarter of its original value, what is the final pressure of the gas? Has the rms speed of the gas molecules increased, decreased, or stayed the same?

31. An ideal gas has the following conditions: $V_i = 400\ \text{cm}^3$, $P_i = 1$ atm, and $T_i = 127$°C. If the pressure is increased to 3 atm and the volume is reduced to 100 cm^3, what is the final temperature of the gas?

32. An ideal gas has an initial volume of 500 cm^3, an initial temperature of 20°C, and an initial pressure of 2 atm. What is its final pressure if the volume is allowed to expand to 1000 cm^3 while the temperature increases to 60°C?

*33. A tank of helium has a volume of 4 L and a pressure of 80 atm. How many balloons can it fill if each balloon has a volume of 1 L and requires a pressure of 1.1 atm?

*34. A 0.6 m^3 tank of helium has a pressure of 100 atm. Each balloon requires a pressure of 1.25 atm and has a volume of 5000 cm^3. How many balloons can be filled from the tank?

12 — STATES OF MATTER

12.1 Density

Density is an important, fundamental property of matter because it is an *intrinsic* property. This means that density doesn't change with a change in the amount of material or the shape of the object. A gold wedding ring has the same density as a pound of gold, and the pound of gold could be cubical, spherical, or any other shape. However, the object cannot be hollow. The density of a hollow object is just the average density of the material and the air inside. Composite materials have a range of densities depending on the relative amounts of the different substances.

The units of density in the SI system are kg/m^3. However, the units g/cm^3 are also widely used. Convince yourself that you can convert to g/cm^3 by dividing the density in kg/m^3 by 1000. Therefore, the density of water is 1000 kg/m^3 = 1 g/cm^3. The densities of some common substances are given in Table 12-1 in the text.

Example 12.1.1

What is the average density of the Sun?

The mass and radius of the Sun are given in the tables in the Appendix. Because the Sun is very nearly spherical, we can calculate its volume V_S to be

$$V_S = \tfrac{4}{3}\pi r^3 = (4.19)\left(6.96\times10^8\,\text{m}\right)^3 = 1.41\times10^{27}\,\text{m}^3$$

The Sun's density is therefore

$$D_S = \frac{M_S}{V_S} = \frac{1.99\times10^{30}\ \text{kg}}{1.41\times10^{27}\ \text{m}^3} = 1410\ \text{kg/m}^3$$

Therefore, the average density of the Sun is 1.4 times that of water. This figure is an average density for the Sun. The density varies from a high of 150 g/cm^3 at its center to a low of 2.3×10^{-8} g/cm^3 at its visible surface. The average density of Earth is 5.5 g/cm^3.

Practice: If the mass of the Sun was compressed to a volume equal to that of Earth, what would be the new density of this compact Sun?
Answer: 1.84×10^9 kg/m^3

Example 12.1.2.

A ball of solid gold hangs from a string. If the tension in the string is 200 N, find the radius of the ball.

The free-body diagram for the golden ball has only two forces acting on it: the gravitational force down and the tension up. The two forces must balance by Newton's second law, so the mass of the ball is given by:

$$m = \frac{W}{g} = \frac{200 \text{ N}}{9.8 \text{ m/s}^2} = 20.4 \text{ kg}$$

The density of gold is known to be 19,300 kg/m^3, so we can use the mass to calculate the volume of the ball:

$$V = \frac{m}{D} = \frac{20.4 \text{ kg}}{19{,}300 \text{ kg/m}^3} = 1.06 \times 10^{-3} \text{ m}^3$$

The radius of the ball can now be calculated using the formula for the volume of a sphere:

$$V = \tfrac{4}{3}\pi r^3 \quad \Rightarrow \quad r = \sqrt[3]{\frac{3V}{4\pi}} = \sqrt[3]{\frac{3\left(1.06 \times 10^{-3} \text{ m}^3\right)}{12.6}} = 6.33 \times 10^{-2} \text{ m} = 6.33 \text{ cm}$$

12.2 Elasticity

We started our discussion of solids by stating that they are rigid; unlike liquids and gases, solids have a definite shape. This is only relatively true. One of the primary reasons early civilizations used gold and silver in their jewelry was its ability to be easily shaped by the jeweler, not its inherent monetary value.

All materials distort when subjected to external forces. The amount of distortion depends on the size of the external force and the type of bonding in the solid. Ionic crystals, for example, distort less than metals. When a force tries to compress or stretch many materials, the change in their size is too small to observe directly. Rubber bands, silly putty, and springs are obvious exceptions to this. If a rubber band or spring is stretched, it will return to its original shape provided the stretch is not too large. Under these conditions, the object obeys Hooke's law

$$F = kx$$

where F is the net force on the spring or rubber band and x is the displacement, or the change in length, and k is the *spring constant.* The spring constant tells us how much force is required to stretch (or compress) the spring a unit length. In the SI system of units, the spring constant is usually expressed in newtons per meter (N/m).

Example 12.2.1

A bathroom scale, which is spring loaded, is compressed 3 cm when an 80-kg person stands on it. What's the spring constant for this bathroom scale?

$$k = \frac{F}{x} = \frac{mg}{x} = \frac{(80 \text{ kg})\left(9.8 \text{ m/s}^2\right)}{0.03 \text{ m}} = 26{,}100 \text{ N/m}$$

Practice: A bathroom scale is being designed by an engineering student. She wants the scale's spring to compress only 3 cm when a 115-kg person stands on it. What spring constant is needed?
Answer: 37,600 N/m

12.3 Pressure

In the section in the text entitled *Pressure* we said that pressure increases as one goes deeper into a liquid. We justified this statement by showing that the pressure at that level must be large enough to support a column of liquid above this level. As one goes deeper, the column gets longer and the pressure required to support the weight of the column must increase. The upward force is the product of the pressure P and the cross-sectional area A of the column. The weight W of the column is Mg where M is the column's mass. Equating these two expressions, we have

$$PA = W = Mg$$

The mass can be obtained from the density D and the volume V of the liquid.

$$M = DV = DAh$$

where h is the height of the column, or the depth of the level. Substituting this expression for the mass into our equation for the pressure and canceling the areas, we obtain the relationship for pressure.

$$P = Dgh$$

If the liquid exists in the atmosphere, we need to add the atmospheric pressure to find the total pressure at a depth h in the liquid.

This relationship also holds for the atmosphere, but it is more complicated because the density varies with altitude, and the top of the atmosphere is hard to determine. Solids and liquids are easier because they are nearly incompressible, making density variations with depth negligible.

The atmospheric pressure at a particular level can be measured with a barometer. The atmospheric pressure is equal to the pressure at the bottom of the mercury column because the air pressure supports the weight of the mercury column.

Example 12.3.1

What is the height of a mercury column if the air pressure is equal to 1 atm? The density of mercury is 13,600 kg/m^3 and 1 atm = 1.01×10^5 N/m^2.

$$h = \frac{P}{Dg} = \frac{1.01\times10^5 \text{ N/m}}{\left(13{,}600 \text{ kg/m}^3\right)\left(9.8 \text{ m/s}^2\right)} = 0.758 \text{ m}$$

Example 12.3.2

If you made a barometer with water instead of mercury, how high would the column of water reach?

The pressures at the bottom of the barometers are both equal to the atmospheric pressure, so we can set them equal to each other.

$$D_w g h_w = D_m g h_m$$

where the subscripts w and m refer to water and mercury, respectively. Canceling the common factor g and solving for the height of the water column, we have

$$h_w = h_m \left(\frac{D_m}{D_w} \right) = 0.758 \text{ m} \left(\frac{13.6 \text{ g/cm}^3}{1 \text{ g/cm}^3} \right) = 10.3 \text{ m}$$

Because mercury is 13.6 times more dense than water, the height of the water column is 13.6 times taller than the mercury column for all values of atmospheric pressure.

Practice: Suppose you made a barometer of oil that has a density of 900 kg/m³. How high would the oil column be when the mercury column is 76 cm tall?
Answer: 11.5 m

12.4 Sink and Float

Archimedes' principle states that an object in a fluid (either a liquid or a gas) is buoyed by a force equal to the weight of the volume of fluid displaced. If the object is suspended from a string while it is completely submerged, as shown in Figure 12.4.1, this buoyant force F_b reduces the force needed to support the object. The force F_s of the string is the weight of the object in the fluid and can be calculated from

$$F_s = W - F_b = D_o V_o g - D_f V_o g = (D_o - D_f) V_o g$$

where V_o is the volume of the object, g is the acceleration due to gravity, and D_o and D_f are the densities of the object and fluid, respectively. How would this equation change if the object were less dense than the fluid and anchored to the bottom by the string?

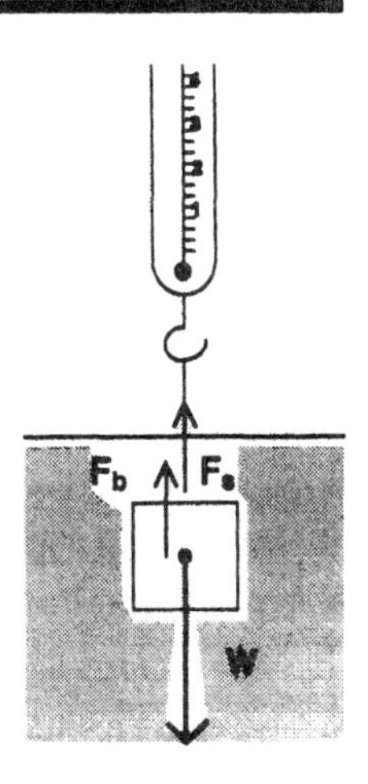

Fig. 12.4.1

Example 12.4.1

A sphere of aluminum has a volume of 0.01 m³ and a density of 2700 kg/m³. In air it weighs 265 N. What is the weight of the sphere when it is under water?

$$W = (D_o - D_f) V_o g$$

$$= \left(2700 \text{ kg/m}^3 - 1000 \text{ kg/m}^3\right)\left(0.01 \text{ m}^3\right)\left(9.8 \text{ m/s}^2\right) = 167 \text{ N}$$

How much of an object floats above the water depends on its density. Because the density of clay is greater than that of water, we know that an equal volume of water will weigh less than the clay. Thus, the clay will sink, as we expect. We can make the clay a "floater" by molding it into the shape of a small bowl. When we set it onto the water's surface, it starts to sink, displacing water. It continues to sink until it has displaced a volume of water that weighs the same as the clay. Under this condition

$$D_o V_o = D_f V_f$$

where V_f is the volume of the water displaced, D_o and V_o are the average density and volume of the bowl. The average density of the bowl is less than the density of solid clay. The product of the average density and the volume of the clay boat is equal to the mass of the clay.

Example 12.4.2

How far will a plastic cup sink in water? Assume that the cup is a cylinder with a cross-sectional area of 6 cm^2, that it has a mass of 30 g, and that it remains vertical.

The cup will sink until it has displaced 30 g of water. Because the density of water is 1 g/cm^3, it will sink until it displaces 30 cm^3 of water.

$$h = \frac{V}{A} = \frac{30 \text{ cm}^3}{6 \text{ cm}^2} = 5 \text{ cm}$$

Practice: If a 10-g mass is placed in the cup, how much farther will the cup sink?
Answer: An additional 1.67 cm

Example 12.4.3

A box with sides 20 cm long is floating in water (ρ = 1000 kg/m^3). Lead balls are added such that the box is one-half submerged (that is, the top of the box is 10 cm above the surface of the water). Determine the total weight of the box (including the lead balls).

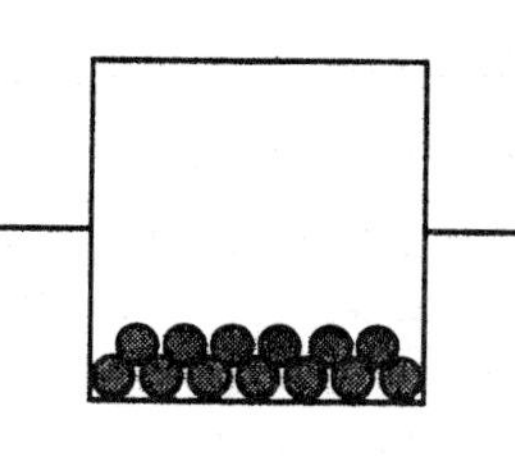

A free-body diagram of the box has only two forces acting on it: the gravitational force acting down and the buoyant force acting up. These two forces must cancel by Newton's second law. The buoyant force is also equal to the weight of the water displaced (by Archimedes' principle):

$$W_{\text{box}} = W_{\text{water displaced}} = D_{\text{water}} V_{\text{water displaced}} g$$
$$= \left(1000 \text{ kg/m}^3\right)\left(4\times10^{-3} \text{ m}^3\right)\left(9.8 \text{ N/kg}\right) = 39.2 \text{ N}$$

Practice: How much more mass could be added to the box before the box sinks?
Answer: 4 kg.

Problems

1. What is the density of cottonseed oil if 1 L has a mass of 926 g?
2. What is the density of ethyl alcohol if 2 L has a mass of 1.58 kg?
3. What is the average density of Saturn? If Saturn could be placed in a truly astronomical bathtub, would it sink or float?
4. A neutron star has about 1.4 times the mass of the Sun and a radius of about 16 km. What is the average density of a neutron star?
5. You have a solid ring that you suspect is pure gold. If you measure the ring to have a volume of 0.05 cm^3, what should the ring's mass be?
6. If you have a solid cube of aluminum that you measure to be 3 cm on each edge, how much mass will you have?
7. What volume does 1 kg of mercury occupy?
8. What is the volume of a lead weight that has a mass of 20 g?
9. A 1-kg block hangs from a spring causing the spring to stretch 6 cm. What is the spring constant in N/m?
10. If a 15-kg mass compresses a scale 0.3 m, what is the spring constant in N/m?
11. A spring has a spring constant of 400 N/m. If a 1.5-kg mass is hung from it, how far will it stretch?
12. How far will a spring with a spring constant of 500 N/m stretch if a 3 kg mass is hung from it?

*13. Two springs with spring constant k = 100 N/m are hung end to end. If a block of mass 4 kg is hung from the end, how much will each spring stretch? What is the effective spring constant of the pair?

*14. Two springs with spring constant k are hanging side by side. If a mass is hung from both of them, what is the effective spring constant of the pair?

15. A scuba diver is 12 m below the surface of a lake. What is the pressure on her body just due to the water?
16. What is the pressure at a depth of 1 km under water? Is atmospheric pressure important?
17. How deep must one go in a tank of mercury to reach a pressure of 6 atm?
18. How deep must one go in water to reach a pressure of 6 atm?
19. A dam holds a lake back that has a surface area of 100 km^2 and a depth of 20 m. What is the water pressure at the base of the dam?
20. What is the pressure at the base of a gasoline storage tank that is 12 m in diameter and 6 m tall? The density of gasoline is 0.68 g/cm^3.
21. Scuba divers are always cautioned against holding their breath while ascending. Which is more dangerous, ascending from 30 m to 25 m or from 5 m to the surface with your breath held? *Hint*: Look at the percentage change in pressure.

*22. A scuba diver at a depth of 20 m achieves neutral buoyancy by blowing 4 L of air into his buoyancy compensator device (an air vest). He then swims down an additional 10 m. Use the fact that the product of pressure and volume for a gas is a constant to find the new volume of the air vest. What is the net force on the diver at a depth of 30 m?

23. What is the weight of 1 kg of gold under water?

24. A block of metal has a mass of 2.4 kg and a volume of 0.001 m^3. What is the weight of the block under water?

25. A piece of metal weighs 0.9 N in air and 0.6 N when submerged in water. What is its volume?

26. A piece of wood weighs 30 N. When placed in water, it just barely floats. What is its volume?

*27. A beaker of water, filled to its brim, rests on a scale, which reads 50 N. A block of aluminum is submerged in the beaker and the water that spills is wiped up. The block rests on the bottom of the beaker and the scale now reads 60 N. What is the mass of the aluminum block?

28. A block of aluminum has a mass of 2.7 kg. If it were reshaped into a boat, how much volume would it need to displace in order to float?

*29. A sphere has a mass of 365 g and a volume of 635 cm^3. What fraction of the sphere will float above the surface of the water?

*30. A hollow block of aluminum is floating in water. If it has a mass of 27 kg and a volume of 0.1 m^3, how much of it will be above the surface?

13 — THERMAL ENERGY

13.1 Mechanical Work and Heat

It takes a lot of mechanical work to raise the temperature of something. This is good. If the opposite were true, scuffing our feet on the floor could end up cooking them. In Chapter 7 we learned that the work done on an object is equal to the net force multiplied by the distance through which the force is applied (the force has to be in the direction of the displacement). Therefore, the frictional force associated with scuffing our feet acting through the lengths of the scuffs does mechanical work which is equivalent to heating the object.

Example 13.1.1

Imagine that you are sitting in a bathtub with 60 liters of water and that the water is too cool (40°C). You decide to heat the water by moving your arms back and forth through the water. Under the very unrealistic assumption that there is no heat loss from the water, how long would it take to raise the temperature of the water 1°C?

Let's assume that you move your arms back and forth once every second, each swing is ½-m long, and the average force F exerted is 40 N in each direction. Therefore, the amount of mechanical work W performed each second is the force times the distance d moved.

$$W = Fd = (40\text{ N})(1\text{ m}) = 40\text{ J}$$

Because this is the energy per second, the power P is 40 W.

To calculate the amount of heat needed to raise the temperature of the water, we need to know the mass m of the water.

$$m = DV = (1\text{ g/cm}^3)(60\text{ L})\left[\frac{1000\text{ cm}^3}{1\text{ L}}\right] = 6 \times 10^4\text{ g}$$

Because it takes 1 cal to raise the temperature of 1 g of water by 1°C, the amount of heat Q required to raise the temperature of this water by 1°C is 2×10^4 cal.

Dividing the total amount of thermal energy required by the rate of generating mechanical energy yields the time required. However, we must remember to convert one of the energy units to the other.

$$t = \frac{E}{P} = \frac{6\times10^4\text{ cal}}{40\text{ J/s}}\left(\frac{4.19\text{ J}}{1\text{ cal}}\right) = 6290\text{ s} = 105\text{ min}$$

Practice: How would the time required change if you decided to push twice as hard?
Answer: It would take one-half as long.

13.2 Specific Heat

Toward the end of this section in the text we talk about bringing two objects into thermal contact with each other. If this is done in a perfectly insulated environment or if the time we are considering is very short, we can assume that there is no flow of energy out of the system. This means that the energy lost by the hot object is equal to the energy gained by the cold object. However, the temperature changes of the objects are not necessarily the same.

Example 13.2.1

Consider a 25-g aluminum cube at a temperature of T_a = 60°C. If this cube is placed in 100 g of room-temperature water (T_w = 20°C), what is the final temperature of the system?

We set the heat lost by the aluminum equal to the heat gained by the water. From Table 13-1 in the text the specific heats for water c_w and aluminum c_a are 1 cal/g·°C and 0.215 cal/g·°C, respectively.

$$m_w c_w \left(T_f - T_w\right) = m_a c_a \left(T_a - T_f\right)$$

Solving for the final temperature, we get

$$T_f = \frac{m_w c_w T_w + m_a c_a T_a}{m_w c_w + m_a c_a}$$

$$= \frac{(100\text{ g})(1\text{ cal/g}\cdot°\text{C})(20°\text{C}) + (25\text{ g})(0.215\text{ cal/g}\cdot°\text{C})(60°\text{C})}{(100\text{ g})(1\text{ cal/g}\cdot°\text{C}) + (25\text{ g})(0.215\text{ cal/g}\cdot°\text{C})} = 22°\text{C}$$

Practice: If the cube were made of copper (c_c = 0.092 cal/g·°C), what would the final temperature be?
Answer: 20.9°C

In the feature *How Fatty Are You?* in Chapter 12 of the text, we learned how Archimedes could have used the intrinsic property of density to determine if King Hieron's crown was made of pure gold. Archimedes could have also solved the mystery by using the intrinsic property of specific heat.

Example 13.2.2

Assume that King Hieron's crown has a mass of 1 kg. Use the data in Table 13-1 of the text to find the heat capacity of the crown (a) if it is pure gold and (b) if it is half gold and half silver.

If the crown is pure gold, the heat capacity of the crown is the specific heat of gold multiplied by the mass of the crown.

$$C_g = c_g m = (0.031\text{ cal/g}\cdot°\text{C})(1000\text{ g}) = 31\text{ cal/°C}$$

If the artisan used silver in making the crown, the heat capacity is

$$C_{gs} = c_g m_g + c_s m_s = (0.031 \text{ cal/g}\cdot°\text{C})(500 \text{ g}) + (0.057 \text{ cal/g}\cdot°\text{C})(500 \text{ g})$$
$$= 44 \text{ cal/°C}$$

Archimedes could have heated the crown by placing it in boiling water. By placing the heated crown in room-temperature water and observing the final temperature of the water, he could tell if the crown were made of pure gold. (Of course, he would have first needed to invent the concepts of energy, heat, and temperature because he lived around 250 BC.

Practice: Find the heat capacity of the crown if it is 25% gold and 75% silver.
Answer: 50.5 cal/°C

13.3 Change of State

The latent heat for a substance is the amount of energy required to change the physical state of a unit mass. For example, when water changes from a solid (ice) to a liquid, 80 cal/g (334 kJ/kg) are required to break the bonds in the ice. The values for the latent heat given in Table 13-2 in the text show that the values vary over a large range. In the reverse process the same amount of energy is given off. Therefore, this table also gives the latent heat of solidification and condensation.

Example 13.3.1

A 20-kg block of ice at 0°C melts to form a puddle of water at 0°C. How much energy is absorbed by the ice?

The latent heat of melting for water is 334 kJ/kg, so the energy needed to change the physical state of the block from ice to liquid water is

$$Q = m_w L_w = (20 \text{ kg})(334 \text{ kJ/kg}) = 6680 \text{ kJ}$$

Although the puddle is still at the same temperature, it has more energy. The energy went into breaking the molecular bonds.

Practice: If there was half as much ice, how much energy would be absorbed?
Answer: 3340 kJ

Example 13.3.2

If the process is reversed; that is, the puddle at 0°C is frozen, how much energy will be given off by the water?

The same amount of energy will be given off. This is very much like a spring that has been stretched and, at some later time, will be released. The energy is stored in the spring and then released.

13.4 Conduction

Heat is conducted in a material through exchanges of kinetic energy between its molecules. Conduction occurs when there are temperature differences in the material. The rate Q/t at which heat passes through a slab of material depends on the temperature difference ΔT between the two sides, the thermal conductivity k of the material, and the area A and thickness L of the slab. This can be written as

$$\frac{Q}{t} = k\frac{A\Delta T}{L}$$

Example 13.4.1

An iron plate is 2 cm thick with a cross-sectional area of 300 cm². One face is held at 150°C and the other is 10°C colder. How much heat passes through the plate each second? Assume that the thermal conductivity for iron is 80 W/m·°C.

$$\frac{Q}{t} = k\frac{A\Delta T}{L} = \left(80\frac{\text{W}}{\text{m}\cdot{}^\circ\text{C}}\right)\frac{(300\ \text{cm}^2)(10^\circ\text{C})}{2\ \text{cm}}\left(\frac{1\ \text{m}}{100\ \text{cm}}\right) = 1200\ \text{W}$$

Practice: What happens to the rate if two plates are placed face to face?
Answer: The rate is cut in half.

Example 13.4.2

A single pane, sliding glass door that has an area of 2 m² and thickness of 0.8 cm. How much heat is lost each second through the door if the temperature inside the house is 20°C and the temperature outside is –10°C?

$$\frac{Q}{t} = k\frac{A\Delta T}{L} = \left(0.8\ \frac{\text{W}}{\text{m}\cdot{}^\circ\text{C}}\right)\frac{(2\ \text{m}^2)(30^\circ\text{C})}{0.008\ \text{m}} = 6\ \text{kW}$$

13.5 Radiation

All objects emit and absorb electromagnetic radiation. We'll discuss electromagnetic radiation more fully in Chapter 22. For now you need to know that this radiation takes the form of waves, which have a wavelength and a frequency. The rate P at which an object emits radiation depends on its area A and its absolute temperature T as given by the Stefan-Boltzmann law.

$$P = e\sigma AT^4$$

where e is the emissivity and σ is Stefan's constant, which has the value $\sigma = 5.67 \times 10^{-8}\ \text{W/m}^2\cdot\text{K}^4$ for all substances. The emissivity ranges in value from 0 to 1 and depends on the surface characteristics of the object. An emissivity of 1 means that the object is a perfect absorber of radiation; all of the radiation incident on the object is absorbed. An emissivity of 0 means that the object is a perfect reflector of radiation; none of the radiation is absorbed.

The rate at which an object absorbs radiation is given by the same formula with the temperature T_o of the surroundings substituted for T. Therefore, the net power radiated by the object is given by

$$P = e\sigma A\left(T^4 - T_o^4\right)$$

If the object is warmer than the surroundings, the object will have a net energy loss and its temperature will drop. If the surroundings are warmer, the object will have a net gain and its temperature will rise. This difference between the rates of emission and absorption will exist until an equilibrium is reached. This is the reason a room can feel cold even after the furnace has warmed the air. The walls are still colder than the air and there is a net flow of radiation to the walls.

Example 13.5.1

What is the net rate at which a naked person radiates energy in a room at 20°C?

Let's assume that the person approximates a perfect emitter (and absorber) with an emissivity $e = 1$ and has a surface area of 1.5 m^2. Skin temperature is usually a few degrees below body temperature of 37°C, so let's assume a skin temperature of 33°C = 306 K.

$$P = e\sigma A\left(T^4 - T_o^4\right)$$

$$= (1)\left(5.67 \times 10^{-8}\,\frac{\text{W}}{\text{m}^2\cdot\text{K}^4}\right)\left(1.5\ \text{m}^2\right)\left[\ (360\ \text{K})^4 - (293\ \text{K})^4\ \right] = 119\ \text{W}$$

This is a large power loss, a little more than a 100-W light bulb. A person would need to consume at a rate of about 2500 Calories/day to make up for this loss. Remember that calories with a capital C means food calories, or kcal.

The electromagnetic radiation emitted and absorbed by all objects has a range of frequencies and wavelengths. The distribution of wavelengths—the intensity of the radiation in each wavelength region—depends on the object's temperature. The distribution drops to zero at high and low wavelengths and has a characteristic hump at intermediate wavelengths as shown in Figure 23-14 in the text. The wavelength λ_{max} corresponding to the peak intensity is given by Wien's law.

$$\lambda_{max} = \frac{2.9\times10^6\ \text{nm}\cdot\text{K}}{T}$$

where the temperature T must be expressed in kelvin.

Example 13.5.2

What is the peak wavelength radiated from your skin?

$$\lambda_{max} = \frac{2.9 \times 10^6 \text{ nm} \cdot \text{K}}{T} = \frac{2.9 \times 10^6 \text{ nm} \cdot \text{K}}{306 \text{ K}} = 9480 \text{ nm}$$

According to Figure 22-26 in the text, this wavelength lies around the middle of the infrared region, an invisible part of the electromagnetic spectrum. If we were to shut off all the lights in a room, you would not be visible with human eyes. Infrared sensing devices, however, would "see" you.

Practice: What is the peak wavelength of a stove element at a temperature of 800°C?
Answer: 2700 nm. Although this peak wavelength is still in the infrared, we can see the element glow red because the wavelength distribution has enough contribution to the visible region of 400-750 nm.

13.6 Thermal Expansion

We argued in the text that the thermal expansion ΔL in the length of an object depended on the original length L, the change in temperature ΔT, and the coefficient of thermal expansion α, which is characteristic of the material.

$$\Delta L = \alpha L \Delta T$$

Example 13.6.1

If an aluminum bar has a length of 10.0000 m at a temperature of 20°C and a length of 10.0096 m at 60°C, what is the coefficient of thermal expansion of aluminum?

Solving our equation for α, we have

$$\alpha = \frac{\Delta L}{L \Delta T} = \frac{0.0096 \text{ m}}{(10.0000 \text{ m})(40°\text{C})} = 2.4 \times 10^{-5}/°\text{C}$$

Problems

1. While a system is being cooled, 180 J of heat are removed and 90 J of work are done on it. What is the change in the internal energy of the system?

2. If it requires 400 J of work to compress an ideal gas and its internal energy increases by 150 J, how much heat is released?

3. A force of 35 N is required to push a crate at constant speed across the floor. How far does it have to be pushed to generate 1000 calories of thermal energy?

4. A 1-kg book falls from a table that is 70 cm high. How many calories of thermal energy are produced?

*5. In a laboratory exercise to verify Joule's result for the mechanical equivalent of heat, a student observed that when a 45-kg mass dropped a distance of 2 m, the temperature of 1000 g of water rose by 0.2°C. How many joules are equivalent to one calorie? How might you account for the difference in this result and the one given in the chapter?

6. It is often said that a person gives off heat equivalent to a 75-W light bulb. How many Cal/s is this?

7. A typical daily allowance of food energy is about 2500 kcal. If a 50-kg mountaineer could convert food energy completely to gravitational potential energy, what is the tallest mountain she could climb while "burning" 2500 kcal?

8. How many Cal/s would need to be consumed to equal the power of a 1500-W hair dryer?

9. How much heat is released when 1 kg of water is cooled from 20°C to 0°C?

10. How much heat is need to raise the temperature of 1 kg of water by 5°C?

11. How much heat is required to change the temperature of 1 kg of copper by 50°C?

12. Assume that you measure the heat required to change the temperature of a fixed mass of water by 5°C. What temperature change will occur if you supply the same amount of heat to an equal mass of gold?

*13. A 100-g sample of an unknown metal at 50°C is dropped into an insulated beaker containing 50 g of water at 10°C. When the system reaches equilibrium, the temperature is 22°C. What is the specific heat of the metal?

14. The temperature of an unknown metal increases by 50°C when 92 cal of heat are added. If the mass of the metal is determined to be 20 g, identify a possible metal from the data in Table 13-1.

*15. One hundred grams of gold at 70°C is placed into a completely insulated container containing an equal mass of water at 30°C. If the container is made of aluminum and has a mass of 10 g, calculate the final temperature of the gold-water-aluminum system.

16. An unknown amount of water at 100°C is mixed with 200 g of water at 50°C in a completely insulated container. If the final equilibrium temperature of the mixture is 80°C, what is the mass of the hot water?

17. Fifty grams of gold at a temperature of 100°C are placed in 100 g of water at 19°C. What is the equilibrium temperature if no energy is lost to the surroundings?

18. What is the equilibrium temperature if a 200-g sample of aluminum at 80°C is placed in 150 g of water at 18°C?

19. When a sample of copper at 96°C is placed into 200 g of water at 20°C, the equilibrium temperature is 22°C. What is the mass of the sample?

*20. Given that 1 lb. of water has a mass of 454 g, show that 1 BTU is equal to 252 cal.

21. A statue of a cowboy made of 4 kg of copper except for his hat, which is made of 200 g of silver. What is the heat capacity of the statue?

22. When your authors were children, dental fillings were made of a mixture of 50% silver and 50% mercury. What is the heat capacity of a 150-g filling?

23. How much heat is released when 500 g of water at 0°C freezes?

24. How much heat is required to convert 500 g of water at 100°C to steam at 100°C?

25. How does the amount of heat required to convert 1 g of water at 100°C to steam at 100°C compare to the amount of thermal energy required to heat 1 g of water from 20°C to 100°C? (This is why some radiators use steam rather than hot water.)

26. How much heat is required to convert 500 g of ice at −10°C to water at +10°C?

*27. What is the maximum mass of 100°C water that can be cooled to 0°C with 100 g of ice at 0°C?

28. A 20-g ice cube at 0°C is placed in a 50-g cup of water at 50°C. Will the ice cube completely melt?

29. A concrete wall is 10 cm thick and has a surface area of 5 m^2. At what rate will thermal energy pass through the wall if the inside and outside temperatures are 20°C and −10°C?

30. A 1.9-cm thick wood wall with a surface area of 5 m^2 has a temperature of 20°C on one side and 0°C on the other. What is the heat loss through the wall per second?

31. If Earth were a perfect radiator (it isn't), it would have a temperature of 400 K. If this were true, how much energy would Earth radiate for each square meter of its surface?

*32. What is the power radiated by the Sun, assuming that it is a perfect radiator with a temperature of 5800 K?

33. Assuming a skin temperature of 30°C, what is the peak wavelength of the radiation emitted from your body?

34. What temperature is required for an object to have a peak wavelength in the middle of the visible spectrum?

35. What is the peak wavelength emitted by the Sun? Assume that the surface temperature of the Sun is 5800 K.

36. Why might a large star with a surface temperature of 3000 K be called a red giant?

37. If telephone wires between poles separated by 60 m are fairly taut at a temperature of −30°C, how much longer will the wire be when the temperature reaches +40°C? Copper has a coefficient of thermal expansion of 1.7×10^{-5}/°C.

38. If sections of a concrete highway are 20 m long when poured at a temperature of 20°C, how big a space must be left between sections if the concrete could reach temperatures of 50°C? The thermal coefficient of thermal expansion for concrete is 1.2×10^{-5}/°C.

39. A 20-cm diameter hole is cut in a steel plate at 20°C. What is the diameter of the hole when the temperature drops to −20°C?

40. If steel rails used by the Union Pacific Railroad have a length of 30 m and a gap between rails of 2 mm at 20°C, what size gap will appear at −30°C?

*41. The volume coefficient of thermal expansion is 1.82×10^{-4}/°C for mercury. If the volume of mercury is 1 liter at 20°C, what is its volume at 40°C?

*42. An aluminum plate is 10 cm by 10 cm at 20°C. By how much will its area increase when it is heated to 100°C?

14 — AVAILABLE ENERGY

14.1 Real Engines

The maximum efficiency of your automobile's engine can be calculated using Carnot's relationship for an ideal heat engine. The Carnot efficiency η_c is given by

$$\eta_c = 1 - \frac{T_c}{T_h}$$

where T_h and T_c are the absolute temperatures of the hot and cold reservoirs, respectively. To obtain the actual efficiency η, we need to measure the amount of heat put into the engine and the amount of work that it does.

$$\eta = \frac{W}{Q}$$

Example 14.1.1

An engine with a power output of 200 kW operates between temperatures of 500°C and 30°C. If it requires energy at a rate of 6×10^5 J/s to operate, what are its Carnot and actual efficiencies?

To find the Carnot efficiency, we need to convert the Celsius temperatures to the absolute scale and insert them into the Carnot relationship.

$$\eta_c = 1 - \frac{T_c}{T_h} = 1 - \frac{303\text{K}}{773\text{K}} = 60.8\%$$

To calculate the actual efficiency, we look at the energy input and work output during 1 s.

$$\eta = \frac{W}{Q} = \frac{2 \times 10^5\ \text{J}}{6 \times 10^5\ \text{J}} = 33.3\%$$

Therefore, the losses due to friction and the flow of thermal energy to the surroundings reduce the efficiency by 27.5%.

Practice: What are the efficiencies if the higher temperature increases to 600°C?
Answer: η_c = 65.3%; η = 33.3% (no change)

Example 14.1.2

Suppose a gasoline engine has a maximum theoretical efficiency of 30% and you measure the temperature of the exhausted gases to be 30°C. What is the minimum possible temperature inside the cylinders after the combustion?

The hotter the input temperature for a fixed exhaust temperature, the lower the efficiency. Therefore, the minimum temperature is given by the Carnot efficiency.

$$\eta_c = 1 - \frac{T_c}{T_h}$$

Solving for T_h we obtain

$$T_h = \frac{T_c}{1 - \eta_c} = \frac{303\text{ K}}{0.70} = 433\text{ K} = 160°\text{C}$$

A real engine would require a higher temperature to compensate for the lower efficiency.

Practice: If the efficiency of the engine is increased to 40%, what is the new cylinder temperature?
Answer: 505 K = 232°C

14.2 Refrigerators

The "efficiency" of a refrigerator is called the *coefficient of performance* (*COP*); it is equal to the heat Q_c extracted from the cold-temperature region divided by the work W required to make the transfer (see Figure 14-9 in the text).

$$COP = \frac{Q_c}{W}$$

A good refrigerator extracts a lot of heat from the objects that you want to keep cold with a minimum amount of work. Good refrigerators have *COP*'s of 5 or 6.

If we consider our refrigerator to be a Carnot engine running backward, we can get an expression for the maximum *COP* in terms of the temperatures T_h and T_c of the hot and cold regions.

$$COP_c = \frac{T_c}{T_h - T_c}$$

Example 14.2.1

If we have a refrigerator with an internal temperature of 0°C and an exhaust temperature of 33°C, what is its maximum *COP*?

By asking for the maximum *COP*, we are assuming that the refrigerator is a Carnot engine running backward. This means that the efficiency can be expressed as

$$COP_c = \frac{T_c}{T_h - T_c} = \frac{273\text{ K}}{306\text{ K} - 273\text{ K}} = 8.27$$

Practice: What will the exhaust temperature be if you lower the thermostat to –10°C and assume that the *COP* doesn't change?
Answer: 295 K = 22°C

14.3 Entropy

The relationship between the disorder of a system and entropy can be expressed mathematically, as discovered by Ludwig Boltzmann. The Boltzmann equation is

$$S = k \log W$$

where S is the entropy of the system, k is a constant, and W is, loosely speaking, the number of different, but equivalent, ways that a system can be put together. Strictly speaking, this relationship only holds for systems with very large numbers of states. To get an approximate value for log W, use the exponent when W is written in the powers-of-ten notation. You can obtain a better answer by entering the value of W into a calculator and pushing the "log" button.

Example 14.3.1

As an artificial, but illustrative, example let's return to the case of our coins discussed in the text. For simplicity, assume that the constant k is equal to one.

The entropy for the case of all heads (or all tails) would be

$$S = k \log(1) = 0$$

because we can write $1 = 10^0$. The entropy for the case of two heads and two tails would be

$$S = k \log(4) = 0.602$$

because there are four ways of obtaining two heads and two tails. (In this case, we used a calculator.)

In thermodynamics, the system is more disordered when it has a higher number of equivalent states. We see that the Boltzmann relationship shows that the system also has higher entropy.

Practice: In all situations we are interested in the change in the entropy, not the actual value of the entropy. What would be the effect of putting in the correct value of Boltzmann's constant?
Answer: The relative changes would be the same, but the size of the changes would be scaled by the constant.

Problems

1. A heat engine does 300 J of work while exhausting 900 J. What is its efficiency?

2. A heat engine exhausts 400 J of heat each second. If the engine operates at an efficiency of 40%, at what rate is it able to do work?

3. What is the efficiency of a heat engine that takes in 8 cal of heat and does 8.4 J of mechanical work?

4. A salesperson claims that his best engine will do 12 J of mechanical work for each 5 cal of heat taken in. Is this possible?

5. A heat engine takes in 8000 cal of energy at 900 K and exhausts 4500 cal at 400 K. What is the efficiency of this engine? What is its maximum theoretical efficiency?

6. A modern steam turbine typically has a maximum input temperature of 810 K (999°F) and an outlet temperature of 310 K (98°F). What is its maximum theoretical efficiency?

7. An ideal heat engine operates at 50% efficiency at an exhaust temperature of 27 °C. By how much would the input temperature have to be raised to increase the efficiency to 60%?

8. An ideal heat engine operates with an input temperature of 327°C and an exhaust temperature of 27°C. If the input temperature is lowered to 227°C, by how much must the exhaust temperature be lowered to maintain the same efficiency?

*9. A typical coal-fired electric power plant has an efficiency of 38%, while a nuclear power plant is more like 32% efficient. How many joules of thermal energy are required by each plant to generate one joule of electrical energy? How much does each plant exhaust as waste? Is the 6% difference in efficiency important?

*10. A steam turbine has an efficiency of 47%. If one now includes a boiler efficiency of 88% and a generator efficiency of 99%, what is the overall efficiency of the electrical generating system? (*Hint*: 88% of the original energy leaves the boiler, 47% of that leaves the turbine, and 99% of that leaves the generator.)

11. A heat pump has a coefficient of performance of 5. How much heat energy enters the house for every one joule of work done to drive the heat pump?

12. A refrigerator usually exhausts 1200 cal of heat from its internal space with a coefficient of performance of 7. How much work is performed by the refrigerator?

13. A heat pump (working like a refrigerator) extracts energy from a cold reservoir at 0°C and expels it to a hotter reservoir at 20°C. What is its coefficient of performance if you assume that the heat pump is an ideal Carnot engine running backward?

14. An air conditioner moves energy from a room at 20°C and expels it to the outside at 35°C. If you assume that the air conditioner is an ideal Carnot engine running backward, how many joules of heat are removed from the house for each joule of electrical energy used to drive the air conditioner?

15. Suppose you had a paper bag filled with 50 colored marbles. Twenty of the marbles are red and the remainder are white. What is the probability of reaching in and pulling out a white marble?

16. Suppose you have removed a red marble in the previous problem. What is the probability of reaching in and pulling out a second red marble?

17. Some plastic chips have two colors; red on one side and blue on the other. If you put four in a bag, shake the bag, and dump them on a table, how many different ways can you have exactly 2 chips with the red side up?

18. How many different ways can you throw a total of 7 with 3 dice? What is the probability of getting a total of 7 on a single throw?

19. Imagine that you have dice with 12 faces that are all the same so that the likelihood of any one face being on top is the same as any other face. If the faces are numbered 1 through 12, what is the probability of rolling a total of 2 with two such dice?

*20. What is the probability of rolling a total of 13 with the dice in the previous problem?

21. Find the entropy of 100 pennies with all the heads up. There are 1×10^{29} possible arrangements for half heads and half tails. What is the maximum possible change in entropy?

*22. Find the entropy of three identical coins for the different arrangements, assuming that $k = 1$. What is the maximum possible change in entropy?

15 — VIBRATIONS AND WAVES

15.1 Simple Vibrations

For our relationship between period and frequency to work properly, we must be sure that the frequency is stated in cycles per unit time, or an equivalent expression such as revolutions per unit time. [Note that cycles and revolutions are not units in the usual sense that they must be carried through equations.] Sometimes we are given the rotational speed instead of the frequency. In such cases, we must convert these rotational speeds to frequencies in order to calculate the periods.

Example 15.1.1

What are the frequency and period of a disk rotating with a rotational speed of 18°/s?

We need to convert degrees to revolutions. We can do this knowing that 1 rev = 360°.

$$f = \frac{18^\circ}{1\text{ s}}\left[\frac{1\text{ rev}}{360^\circ}\right] = 0.05\text{ rev/s}$$

$$T = \frac{1}{f} = \frac{1}{0.05\text{ rev/s}} = 20\text{ s}$$

Practice: What happens to the period if the frequency is doubled?
Answer: It is half as big.

15.2 Mass on a Spring

Because the relationship for the period T of oscillation of a mass on a spring is given in terms of the mass m and the spring constant k, we can always find one of the variables if given the other two.

Example 15.2.1

What mass is required to yield a period of 1 s when hanging from a spring with a spring constant of 100 N/m?

We begin by squaring both sides of the equation for the period and then solving for the mass. After this we can plug in the given values.

$$T = 2\pi\sqrt{\frac{m}{k}}$$

$$m = \frac{kT^2}{4\pi^2} = \frac{(100 \text{ N/m})(1 \text{ s})^2}{39.5} = 2.53 \text{ kg}$$

Practice: What mass is required to produce a period of 2 s for this spring?
Answer: 10.1 kg

Example 15.2.2

What spring constant would you use with a 1-kg mass to produce a period of 0.5 s?

$$k = \frac{4\pi^2 m}{T^2} = \frac{39.5(1 \text{ kg})}{(0.5 \text{ s})^2} = 158 \text{ N/m}$$

Practice: What spring constant is needed to yield a period of 1 s with this mass?
Answer: 39.5 N/m

When two springs with identical spring constants k are placed side-by-side (or in parallel), we can define an effective spring constant k_{eff} for the combination. Because we have to pull with twice the force to get the same displacement, the spring constant for the combination must be twice as large. In general,

$$k_{eff} = k_1 + k_2$$

Example 15.2.3

The spring constant for a car's suspension system is a result of the combination of four springs. If the spring constant for one of the springs has a value of 10^4 N/m and the car's mass is 1600 kg, with what period will the car oscillate?

Generalizing the case for two springs in parallel, we expect that the effective spring constant will be 4 times the value of one of them, or 4×10^4 N/m. Therefore, the period is given by

$$T = 2\pi\sqrt{\frac{m}{k_{eff}}} = 2\pi\sqrt{\frac{1600 \text{ kg}}{4 \times 10^4 \text{ N/m}}} = \frac{4\pi}{10}\text{s} = 1.26 \text{ s}$$

Practice: A mass on the end of a spring oscillates with a period of 1 s. What will the period be if the mass is attached to two of these springs placed side-by side?
Answer: 0.707 s

We can also consider the case of two springs with identical spring constants placed end-to-end (or in series). Because the applied force is felt by both springs, we get twice the displacement for the same force. Therefore, the effective spring constant is one-half of either spring constant. In general,

$$\frac{1}{k_{eff}} = \frac{1}{k_1} + \frac{1}{k_2}$$

Example 15.2.4

A spring has a spring constant of 10^4 N/m. What happens to the spring constant when you cut the spring in half?

Because this is the opposite of connecting two springs end-to-end, we would expect the spring constant to be twice as large.

$$k_{eff} = 2k = 2 \times 10^4 \text{ N/m}$$

In Section 11.2, we found that the force exerted by a spring is proportional to the stretch or compression of the spring from its equilibrium position. The force is given by Hooke's law.

$$F = kx$$

The work required to compress or stretch a spring a displacement x is the average force times the displacement.

$$W = F_{ave}x = \tfrac{1}{2}\left(F_{min} + F_{max}\right)x = \tfrac{1}{2}\left(0 + kx\right)x = \tfrac{1}{2}kx^2$$

This work is stored as potential energy in the spring that can later be released as the spring does work on something else.

Example 14.2.5

A 4-kg block is attached to a spring and oscillating horizontally on a frictionless table. The spring constant k = 1800 N/m. The maximum speed of the block is v = 10 m/s. When the block is moving at a speed of v = 5 m/s, how far is the spring stretched (or compressed)?

When the speed is a maximum, all of the energy is kinetic. The total energy of the mass-spring system is therefore

$$E_{total} = \tfrac{1}{2}mv_{max}^2 = \tfrac{1}{2}(4\text{ kg})(10\text{ m/s})^2 = 200\text{ J}$$

When the speed is v = 5 m/s, the kinetic energy is

$$E = \tfrac{1}{2}mv^2 = \tfrac{1}{2}(4\text{ kg})(5\text{ m/s})^2 = 50\text{ J}$$

In general the mechanical energy of the system is the sum of the kinetic and potential energy.

$$E_{total} = \tfrac{1}{2}mv^2 + \tfrac{1}{2}kx^2$$

Solving for x, we find the displacement to be

$$x = \sqrt{\frac{E_{total} - \frac{1}{2}mv^2}{\frac{1}{2}k}} = \sqrt{\frac{200\ \text{J} - 50\ \text{J}}{\frac{1}{2}(1800\ \text{N/m})}} = 0.408\ \text{m}$$

Practice: If the block has speed $v = 5$ m/s when the spring has a displacement $x = 0.1$ m, what is the amplitude of the motion?
Answer: 0.256 m

15.3 The Pendulum

Example 15.3.1

Suppose you want to make a clock with a pendulum that has a period of 1 s. How long should the pendulum be?

To solve our relationship for the length, we begin by squaring both sides. We then plug in the given values.

$$T = 2\pi\sqrt{\frac{L}{g}}$$

$$L = \left(\frac{T}{2\pi}\right)^2 g = \left(\frac{1\ \text{s}}{6.28}\right)^2 \left(9.8\ \text{m/s}^2\right) = 0.248\ \text{m}$$

Practice: How long must a pendulum be to have a period of 2 s?
Answer: 0.993 m

Example 15.3.2

The period of a pendulum can be used as an experimental method for determining the value of g. Suppose a student on a distant planet measures the period of a 1-m pendulum to be 3.14 s. What is the value of the acceleration due to gravity on this planet?

Solving the relationship for acceleration due to gravity, we get

$$g = \left[\frac{2\pi}{T}\right]^2 L = \frac{4\pi^2}{(\pi\ \text{s})^2}(1\ \text{m}) = 4\ \text{m/s}^2$$

The formula for the period of a pendulum assumes that the string is massless and that the mass is concentrated at a point. In practice, two pendula with the same length string can have different periods if the bobs have different sizes. Careful study shows that the length of the pendulum is actually the distance from the support point to the center of mass of the bob if the length of the string is large compared to the size of the bob.

If the string has mass, or if the whole object is swinging back and forth, we have to modify our relationship to take this into account. As an example of a more complicated pendulum, consider a meter stick swinging from one end. The period T for such a *physical pendulum* is given by

$$T = 2\pi\sqrt{\frac{I}{mgh}}$$

where I is the rotational inertia of the object about the pivot point, h is the distance from the pivot point to the center of mass of the object, m is the object's mass, and g is the acceleration due to gravity.

Let's check to make sure that this new relationship reduces to the previous result for a simple pendulum. The rotational inertia of a point mass m moving along a circular path of radius r is mr^2 (Section 8.2). Substituting this into the relationship yields

$$T = 2\pi\sqrt{\frac{I}{mgh}} = 2\pi\sqrt{\frac{mr^2}{mgr}} = 2\pi\sqrt{\frac{r}{g}}$$

where r is the length of the pendulum.

As fascinating as what affects the period of the pendulum is what does *not* affect it. Notice that the mass and the amplitude are not in the relationship for the period. The absence of the amplitude means that the period remains constant (at least to a good approximation) as the motion dies down.

Example 15.3.3

What is the period of a meter stick swinging from one end?

We saw in Section 8.2 that the rotational inertia of a rod about one end is

$$I = \tfrac{1}{3}mL^2$$

The value of h is $L/2$ because the center of mass is at the middle of the meterstick. Therefore, the period of the swinging meter stick is

$$T = 2\pi\sqrt{\frac{2L}{3g}} = 2\pi\sqrt{\frac{2\text{ m}}{3\left(9.8\text{ m/s}^2\right)}} = 1.64\text{ s}$$

15.4 One-Dimensional Waves

The speed of a wave can be determined in the same manner as you would determine the speed of a car. You simply measure how far the wave travels in a given amount of time.

Example 15.4.1

If it takes the thunder 11.7 s to reach you from a lightning bolt that strikes a tree 4 km away, what is the speed of sound?

$$v = \frac{d}{t} = \frac{4000 \text{ m}}{11.7 \text{ s}} = 342 \text{ m/s}$$

Practice: What is the speed of a wave that travels down a 20-m rope in 3 s?
Answer: 6.67 m/s

15.5 Periodic Waves

The speed v, wavelength λ, and frequency f of periodic waves are always related by the basic equation derived in the text.

$$v = \lambda f$$

This equation can be rearranged to get the wavelength in terms of the speed and the frequency, or it can be solved for the frequency to obtain its value in terms of the speed and the wavelength.

Example 15.5.1

A periodic wave has a speed of 1500 m/s and a frequency of 50 Hz. What is its wavelength?

$$\lambda = \frac{v}{f} = \frac{1500 \text{ m/s}}{50 \text{ Hz}} = \frac{1500 \text{ m/s}}{50/\text{s}} = 30 \text{ m}$$

Practice: What would the wavelength be if the frequency were increased to 3000 Hz?
Answer: 0.5 m

Example 15.5.2

What is the frequency of a periodic wave with a wavelength of 40 m and a velocity of 5000 m/s?

$$f = \frac{v}{\lambda} = \frac{5000 \text{ m/s}}{40 \text{ m}} = 125 \text{ Hz}$$

15.6 Standing Waves

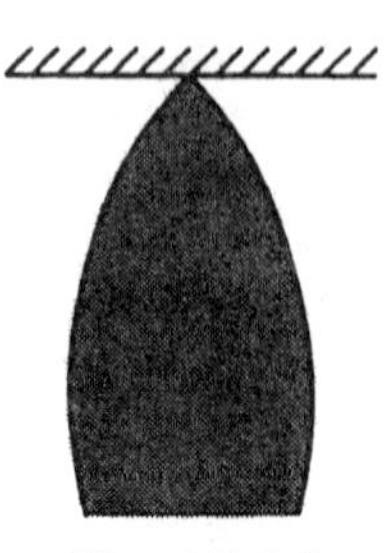
Fig. 15.6.1

The key to determining the possible standing wave patterns is knowing what happens at each end of the medium. For instance, if a rope is fixed at both ends, there will be nodes at each end. However, if the rope is hung vertically, the lower end will be free. Then there will be a node at the upper end and an antinode at the lower end. Because the distance between adjacent nodes and antinodes is one-fourth wavelength, the longest wavelength that will fit on the rope is four times the length of the rope L as shown in Figure 15.6.1. Thus, the fundamental wavelength $\lambda_o = 4L$.

If we now shorten the wavelength, we find that the next pattern that will fit on the rope has a wavelength of $\frac{4}{3}L$, as shown in Figure 15.6.2. Because the wavelength is one-third as long as the fundamental wavelength, this is the third harmonic. The next shorter wavelength is one-fifth λo, and so on. Therefore, only the odd harmonics are present and the pattern for the wavelengths can be written as

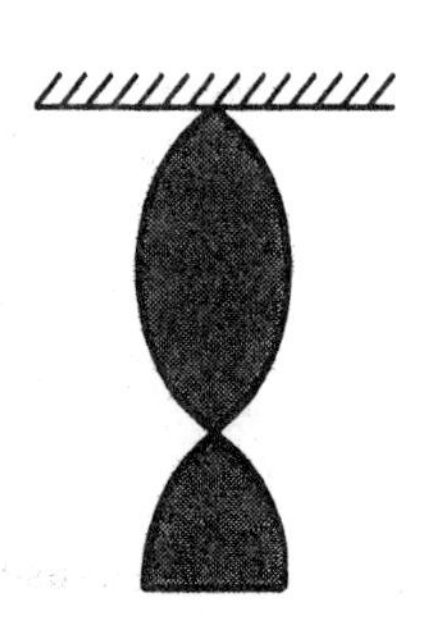
Fig. 15.6.2

$$\lambda = \tfrac{4}{1}L, \tfrac{4}{3}L, \tfrac{4}{5}L, \tfrac{4}{7}L, \ldots$$

Once we know the wavelength of a standing wave, we can calculate its frequency if we know the speed of the waves in the medium. The relationship between speed, wavelength, and frequency for periodic waves holds, because a standing wave is just a superposition of two periodic waves traveling in opposite directions.

Example 15.6.1

What is the frequency of the fundamental standing wave on a 5-m rope fixed at both ends if the waves travel at 2 m/s?

As indicated in the text, the fundamental standing wave will have the longest wavelength. When both ends are fixed, the longest wavelength is twice the length of the rope. Therefore, $\lambda = 10$ m.

$$f = \frac{v}{\lambda} = \frac{2 \text{ m/s}}{10 \text{ m}} = 0.2 \text{ Hz}$$

Practice: What is the frequency of the second harmonic?
Answer: 0.4 Hz

15.7 Interference

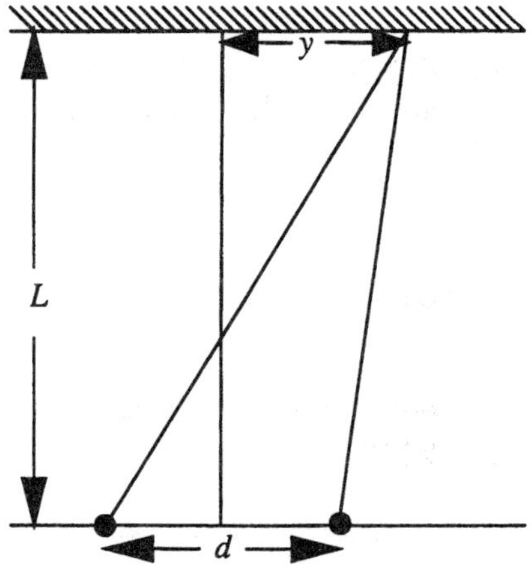

Assume that we have two point sources of water waves that are the same distance away from a distant wall as shown in Figure 15.7.1. If the two sources are in phase, the central antinodal line will be perpendicular to the wall. There are other locations where the waves

Fig. 15.7.1

arrive in phase. In these cases, the wave traveling the greater distance lags behind the other wave by a whole-number of wavelengths. If the distance to the wall is much greater than the source separation and the wavelength, the locations where these antinodal lines intersect the wall can be obtained with the following formula.

$$y_{max} = m\frac{\lambda L}{d} \qquad m = 0, \pm 1, \pm 2, \pm 3, \ldots$$

where y_{max} is the distance measured along the wall from the central bright region, λ is the wavelength, L is the distance to the wall, d is the spacing of the slits, and m is an integer that numbers the antinodal lines beginning with $m = 0$ at the center.

Example 15.7.1

Two sources of water waves with a spacing of 5 cm are 1 m from a wall. If the wavelength of the water waves is 2 cm, what is the spacing of the antinodal lines along the wall?

Because we know that the central antinodal line is at $y = 0$, we can get the spacing by setting $m = 1$.

$$y = m\frac{\lambda L}{d} = (1)\frac{(2\text{ cm})(1\text{ m})}{5\text{ cm}} = 0.4\text{ m}$$

Therefore, the spacing between antinodal regions is 0.4 m. If we were to choose $m = 5$ and $m = 6$, we would get the same answer.

Practice: What is the spacing if the wavelength is reduced to 1 cm?
Answer: 0.2 m

15.8 Diffraction

An expression similar to the one in the last section gives the locations of the nodal positions from a diffraction pattern produced by a single slit of width w.

$$y_{min} = m\frac{\lambda L}{w} \qquad m = \pm 1, \pm 2, \pm 3, \ldots$$

Notice that $m = 0$ is excluded. This occurs because the central region is an antinodal region. This central region is twice as wide as the other antinodal regions.

Example 15.8.1

Waves pass through a 4-cm slit and strike a wall 1 m away. If the wavelength is 2 cm, how wide is the central maximum?

We can calculate the width of the central maximum by determining the locations of the minima on either side. Using our formula with $m = \pm 1$, we have

$$y_{min} = m\frac{\lambda L}{w} = (\pm 1)\frac{(2\text{ cm})(1\text{ m})}{4\text{ cm}} = \pm 0.5\text{ m}$$

Therefore, the central maximum is 1 m wide.

Practice: What is the width of the central maximum if the slit width is increased to 8 cm?
Answer: 0.5 m

Problems

1. A platter rotates with a rotational speed of 30°/s. What are the frequency and period of the platter's motion?
2. If a wheel rotates at 50 radians/s, what are its frequency and period? There are 2π radians in a complete circle.
3. A 1-kg mass is suspended from a spring with a spring constant of 160 N/m. When the mass is at rest, by how much is the spring stretched?
4. A 1-kg mass is suspended from a spring with a spring constant of 160 N/m. What is the system's period of oscillation?
5. If a 0.5-kg block suspended from the end of a spring oscillates with a frequency of 2 Hz, what is the value of the spring constant?
6. For the system in the previous problem, find how much the spring would be stretched with the mass hanging at rest.
7. What mass is needed for a resonant period of 4 s if the spring has a spring constant of 600 N/m?
8. A spring has a spring constant of 400 N/m. What mass is needed to obtain an oscillation with a period of 0.5 s?
9. A 0.50-kg ball is fastened to a spring with a spring constant of 100 N/m and is oscillating horizontally with an amplitude of 0.16 m. What is the maximum potential energy that is stored in the spring during the motion?
10. A 10-kg mass is at rest at the end of an unstretched spring with a spring constant of 4000 N/m. The mass is struck with a hammer giving it a speed of 6.0 m/s. What is the amplitude of the resulting oscillation?
11. A Foucault pendulum has a length of 12 m. What is its period?
12. If a grandfather clock has a pendulum with a length of 1.2 m, what is its period?
13. What would be the length of a simple pendulum with a period of 4 s?
14. A small child on a tire swing completes 4 cycles in 10 s. What is the length of the rope?
15. If a 1-m long pendulum on Mars has a period of 3.26 s, what is the acceleration due to gravity on Mars?
16. If a 1-m long pendulum on the Moon has a period of 4.9 s, what is the acceleration due to gravity near the surface of the Moon?
17. A simple pendulum with a length of 2.0 m has a period of 3.8 s on planet X. What would be the period of a simple pendulum with a length of 8.0 m on this planet?
18. A mass of 2 kg hanging from a spring oscillates with a frequency of 4 Hz. What would

the frequency be if the mass were changed to 8 kg?

*19. A 2-kg block, suspended at the end of a spring, stretches the spring 20 cm. What is the spring constant of the spring? If you were to pull the block down an additional 10 cm and release it, what would be the period of the resulting oscillations?

*20. You hold a dinner plate of radius 20 cm by a point right on its rim. If you pull it back slightly and release it, it oscillates back and forth in simple harmonic motion. Given that the moment of inertia for a disk pivoted about its rim is given by $I = \frac{3}{2}mr^2$, find the period of the oscillations?

21. If 2-Hz waves on a rope travel at 6 m/s, what is their wavelength?

22. If sound waves travel at 340 m/s, what is the wavelength of a musical note with a frequency of 256 Hz?

23. If the fundamental frequency for a 4-m long rope fixed at both ends is 5 Hz, what is the speed of the wave?

24. What is the frequency of the fundamental standing wave produced on a 5-m long rope if the speed of the waves is 8 m/s?

25. What possible wavelengths will produce longitudinal standing waves in a 1-m long rod that is clamped at one end?

26. What possible wavelengths will produce longitudinal standing waves in a 1-m long rod that is clamped at its center?

27. If the speed of sound is 343 m/s, what is the lowest frequency sound that will form a standing wave in a 2-m long tube with antinodes at both ends?

28. If the speed of sound is 343 m/s, what is the lowest frequency sound that will form a standing wave in a 2-m long tube with a node at one end and an antinode at the other?

29. The two probes in a rectangular ripple tank produce waves with a wavelength of 0.5 cm. If the separation of the probes is 2 cm and the far side of the tank is 50 cm away, where would you expect the largest waves to hit the side?

30. Two audio speakers separated by 1 m produce sound waves with a wavelength of 0.5 m. At what points along a facing wall do you expect to hear loud sounds if the wall is 10 m away and the speakers are in phase?

31. Sound waves with a frequency of 1000 Hz and a speed of 343 m/s are incident on a doorway with a width of 80 cm. What is the width of the central maximum on a wall opposite the doorway if the wall is 4 m away?

32. Ocean waves with a wavelength of 10 m and a frequency of 0.2 s strike an opening (width = 10 m) in a seawall straight on. If a flat beach is parallel to the seawall and 200 m from it, where on the beach will the water flow the farthest inland?

16 — SOUND AND MUSIC

16.1 Speed of Sound

The speed of sound in a gas depends on the temperature of the gas and the type of gas. Because we will usually be dealing with the speed of sound in air, a mixture of gases, we will only take a look at the temperature dependence. Experimental measurements and theoretical arguments both agree that the speed of sound in air at other temperatures can be calculated from

$$v = v_o \sqrt{\frac{T}{T_o}}$$

where v_o is the speed at some reference temperature T_o and v is the speed at temperature T. It is necessary to use Kelvin temperatures in this formula. Let's use the values given in the text as our reference values, that is, $v_o = 343$ m/s when $T_o = 20°\text{C} = 293$ K.

Example 16.1.1

What is the speed of sound high in the atmosphere where the temperature is −40°C?

Converting the new temperature to Kelvin, we have $T = 273\text{ K} - 40°\text{C} = 233\text{ K}$.

$$v = v_o \sqrt{\frac{T}{T_o}} = (343\text{ m/s})\sqrt{\frac{233}{293}} = 306\text{ m/s}$$

Thus, an aircraft flying in air with this temperature would break the *sound barrier* if it flew faster than 685 mph.

Practice: What is the speed of sound in air at a temperature of 0°C?
Answer: 331 m/s

16.2 Hearing Sounds

As stated in the text, every increase in the intensity of sound by a factor of 10 means an increase in the sound level of 10 dB. To have a definite scale, some reference intensity must be chosen. By convention 0 dB has been chosen to correspond to the threshold of hearing, which has an intensity of 10^{-12} W/m^2. We can then calculate any other sound level β from the relationship

$$\beta = (10\text{ dB})\log\left(\frac{I}{I_o}\right)$$

If the ratio I/I_0 is a simple power of ten such as 10^3, the log function just yields the exponent, 3 in this case. Otherwise, you can still obtain an answer by putting the ratio into a calculator and pushing the "log" button. (*Caution*: The "ln" button and the "log" button are not the same.)

Example 16.2.1

Let's verify the entry for pain (120 db) shown in the feature "Loudest and Softest Sounds" in the text.

Because the sound intensity of a rock concert is about 1 W/m^2,

$$\beta = (10\text{ dB})\log\left(\frac{I}{I_o}\right) = (10\text{ dB})\log\left(\frac{1\text{ W/m}^2}{10^{-12}\text{ W/m}^2}\right)$$

$$= (10\text{ dB})\log\left(10^{12}\right) = (10\text{ dB})\,12 = 120\text{ dB}$$

Practice: What is the sound level for an intensity of 10^3 W/m^2, the sound of a nearby jet taking off?
Answer: 150 dB

16.3 Stringed Instruments

As shown in the text, the possible wavelengths that can form standing waves on a string fixed at both ends are given by

$$\lambda_n = \frac{2L}{n}$$

where λ_n is the wavelength of the nth harmonic and L is the length of the string. We see that the length of the fundamental wavelength ($n = 1$) is just $2L$.

Because we know the relationship between speed, wavelength, and frequency, we can write down the possible frequencies.

$$f_n = \frac{v}{\lambda_n} = n\left(\frac{v}{2L}\right)$$

We see that the possible frequencies are just whole-number multiples of the fundamental frequency, $v/2L$.

Example 16.3.1

The tension in a guitar string with a length of 80 cm is increased until waves travel along it with a speed of 419 m/s. What are the wavelengths and frequencies of the first three harmonics?

$$\lambda_n = \frac{2L}{n} = \frac{160\text{ cm}}{n} = 160\text{ cm}, 80\text{ cm}, 53.3\text{ cm} \qquad \text{for } n = 1, 2, 3$$

$$f_n = n\left(\frac{v}{2L}\right) = n\left(\frac{419\text{ m/s}}{1.6\text{ m}}\right) = n\,(262\text{ Hz})$$
$$= 262\text{ Hz}, 524\text{ Hz}, 786\text{ Hz} \qquad \text{for } n = 1, 2, 3$$

Practice: What are the wavelength and frequency of the fourth harmonic?
Answer: λ = 40 cm; f= 1050 Hz

16.4 Wind Instruments

If a wind instrument is open at both ends, there is an antinode at each end. This gives the same possible wavelengths and frequencies described in the previous section. On the other hand, if one end is open and the other closed, the possible wavelengths and frequencies change. As described in the text, the fundamental wavelength is four times the length of the tube and only the odd-numbered harmonics are possible. Therefore, the possible wavelengths and frequencies are given by

$$\lambda_n = \frac{4L}{n} \qquad \text{where } n \text{ is odd}$$

$$f_n = \frac{v}{\lambda_n} = n\left(\frac{v}{4L}\right) \qquad \text{where } n \text{ is odd}$$

Example 16.4.1

What are the wavelengths and frequencies of the first three harmonics for a closed organ pipe with a length of 60 cm?

Assuming that we can use the room temperature value for the speed of sound, we have

$$\lambda_n = \frac{4L}{n} = \frac{4(0.6\text{ m})}{n} = \frac{2.4\text{ m}}{n} = 2.4\text{ m}, 0.8\text{ m}, 0.48\text{ m} \qquad \text{for } n = 1, 3, 5$$

$$f_n = n\left(\frac{v}{4L}\right) = n\left(\frac{343\text{ m/s}}{2.4\text{ m}}\right) = n\,(143\text{ Hz}) = 143\text{ Hz}, 429\text{ Hz}, 715\text{ Hz} \qquad \text{for } n = 1, 3, 5$$

Practice: What are the wavelength and frequency for the next higher harmonic?
Answer: λ = 0.343 m; f= 1000 Hz

Example 16.4.2

An organ pipe is found to resonate at 225 Hz, 300 Hz, and 375 Hz. It does not resonate at any other frequencies between 225 Hz and 375 Hz but may resonate at other frequencies outside this range. Is this pipe open or closed? What is its fundamental frequency?

The resonant frequencies are 75 Hz apart, so 150 Hz and 75 Hz must also be resonate frequencies. The lowest frequency must be the fundamental frequency. The pipe must be open because all of the harmonics are present, not just the odd ones.

Practice: A second organ pipe is found to resonate at 150 Hz, 250 Hz, and 350 Hz. It does not resonate at any other frequencies between 150 Hz and 350 Hz but may resonate at other frequencies outside this range. Is this pipe open or closed? What is its fundamental frequency?
Answer: Closed with a fundamental frequency of 50 Hz.

Example 16.4.3

The organ pipe in the last example was open at both ends and had a fundamental frequency of 75 Hz (assuming a room temperature of 20°C). What is the length of this pipe?

The wavelength is given by the wave equation.

$$v = \lambda_1 f_1 \quad \Rightarrow \quad \lambda_1 = \frac{v}{f_1} = \frac{343 \text{ m/s}}{75 \text{ Hz}} = 4.57 \text{ m}$$

Because the fundamental wavelength of an open organ pipe is twice the length of the pipe, the length of the pipe is

$$L = \frac{\lambda_1}{2} = 2.29 \text{ m}$$

Practice: An air column, closed at one end and open at the other (a bottle, for instance), has a fundamental frequency for standing waves of 343 Hz. What is the length of the column?
Answer: 0.25 m

16.5 Beats

When two sound waves of different frequencies are traveling together in the same direction, the superposition of the two waves produces beats like those shown in Figure 16-11(c) in the text. We hear a sound with a frequency equal to the average of the two frequencies that varies in amplitude with a frequency that is equal to the difference in the two frequencies. As mentioned in the text, this phenomenon can be used to tune two strings to the same frequency.

Example 16.5.1

When a key on a piano is pressed, the piano tuner hears a beat frequency of 3 Hz. If one of the strings is known to have a frequency of 262 Hz, what is the frequency of the other?

The two strings must differ in frequency by 3 Hz to produce this beat frequency, but we do not know which string has the higher frequency. Therefore, the other string could have a frequency of 259 Hz or 265 Hz.

Example 16.5.2

You want to know the frequency of a tuning fork. You test it by playing it at the same time as a tuning fork with a known frequency of 342 Hz and you hear beats at a rate of 5 per second. You then play it at the same time as one with a known frequency of 349 Hz and the beats are heard at a rate of 12 per second. What is the frequency of the tuning fork?

The first experiment tells you that the unknown tuning fork is either 342 Hz + 5 Hz = 347 Hz or 342 Hz – 5 Hz = 337 Hz. The second experiment tells you that the unknown tuning fork is either 349 Hz + 12 Hz = 361 Hz or 349 Hz – 12 Hz = 337 Hz. The only frequency that is consistent with both experiments is 337 Hz.

Practice: A guitar string produces 4 beats/s when played with a 340-Hz tuning fork and 8 beats/s when played with a 344-Hz tuning fork. What is the frequency of the guitar string?
Answer: 336 Hz

16.6 Doppler Effect

The change in the frequency of sound that occurs when either the source or the observer is moving can be written in terms of the speeds. If the source is moving, we obtain

$$f = f_o\left(\frac{1}{1 \mp \frac{v_s}{v}}\right)$$

where f_o is the frequency when the source is not moving, f is the new frequency, v is the speed of sound, and v_s is the speed of the source. If the source is moving toward the observer, we use the minus sign, shifting the frequency to a higher value as observed experimentally. On the other hand, we use the plus sign when the source is moving away from the observer.

If the source is stationary and the observer is moving, we use a similar expression.

$$f = f_o\left(1 \pm \frac{v_o}{v}\right)$$

where v_o is the speed of the observer. In this case, we choose the plus sign when the observer is moving toward the source to get the higher frequency. Likewise, we choose the minus sign when the observer is moving away from the source.

Finally, if both the source and the observer are moving, we can multiply the expressions within the parentheses to get the combined effect. The expression is simpler to write and to use if we multiply the numerator and denominator by v.

$$f = f_o\left(\frac{v \pm v_o}{v \mp v_s}\right)$$

where the upper signs are used for motions of the source and receiver toward each other. The choice of signs is easy to remember if you remind yourself that motion toward each other produces a higher frequency.

Example 16.6.1

A train whistle has a frequency f_0 = 200 Hz when it is at rest relative to an observer. What is the frequency heard by an observer at rest if the train is traveling toward the observer at 20 m/s? Assume that the speed of sound is 343 m/s on this day.

$$f = f_o\left(\frac{1}{1 \mp \frac{v_s}{v}}\right) = (200\ \text{Hz})\left(\frac{1}{1 - \frac{20\ \text{m/s}}{343\ \text{m/s}}}\right) = 212.4\ \text{Hz}$$

What if the train is standing still and the observer approaches the train at 20 m/s?

$$f = f_o\left(1 \pm \frac{v_o}{v}\right) = (200\ \text{Hz})\left(1 + \frac{20\ \text{m/s}}{343\ \text{m/s}}\right) = 211.7\ \text{Hz}$$

Notice that the new frequencies are not the same for the two cases. In fact, we kept an extra digit in each answer to show you this difference. The difference would be larger if the velocities were closer to the speed of sound.

What is the observed frequency if both the train and the observer are moving toward each other at 20 m/s?

$$f = f_o\left(\frac{v \pm v_o}{v \mp v_s}\right) = (200\ \text{Hz})\left(\frac{343\ \text{m/s} + 20\ \text{m/s}}{343\ \text{m/s} - 20\ \text{m/s}}\right) = 225\ \text{Hz}$$

Practice: Show that you can get this last answer by using the frequency obtained in the first part of the problem as f_0 in the second calculation.

Problems

1. How much faster does sound travel through air when the temperature is 40°C compared to when it is −40°C?
2. How hot does it have to get for the speed of sound to be 353 m/s?

*3. If the speed of sound in iron is 5100 m/s, how much longer will it take sound to travel 1 km in air compared to 1 km in an iron rail?

*4. What is the time delay between signals traveling in air and water over a distance of 0.5 km?

*5. As you clap your hands at a frequency of 2 Hz, you back away from a wall until the echo reaches you just as you clap the next time. How far from the wall are you?

*6. You watch a carpenter driving nails at a regular rate of one blow per second. The sound of the blows is exactly synchronized with the blows you see. After the carpenter stops hammering, you hear two more blows. How far away is the carpenter?

7. What is the sound intensity in W/m^2 for an 90-dB sound?
8. What is the sound level for a whistle that has a sound intensity of 10^{-3} W/m^2?
9. The sound from a speaker has sound intensity of 10^{-4} W/m^2. By how many decibels does the sound level increase if the intensity is doubled to 2×10^{-4} W/m^2.

*10. Your hand-held sound meter indicates that the sound level from your stereo increases from 75 dB to 80 dB when you push the "loudness" button. By what factor does pushing the "loudness" button increase the intensity?

11. If the speed of the traveling waves on a 60-cm wire is 500 m/s, what is its fundamental frequency?
12. Waves travel at 600 m/s on a wire. If the wire is 75-cm long, what is the fundamental frequency of the wire?
13. If a guitar string has a length of 66 cm and a wave speed of 660 m/s, what are the wavelengths and frequencies of the first three harmonics?
14. If a guitar string has a length of 80 cm and a wave speed of 500 m/s, what are the wavelengths and frequencies of the first three harmonics?
15. What is the fundamental frequency of an open organ pipe with a length of 1 m?
16. How long would an open organ pipe need to be to produce a frequency of 440 Hz?
17. What is the fundamental frequency of a closed organ pipe with a length of 1 m?
18. How long would a closed organ pipe need to be to produce a frequency of 440 Hz?
19. A 2-m tall cylindrical beaker is partially filled with water. On a day when the speed of sound is 340 m/s, the fundamental frequency of this tube is 50 Hz. What is the depth of the water in the bottom of the tube?
20. If a tuning fork of frequency 350 Hz is rung next to the opening of the tube in the previous problem, it sets up standing waves in the

tube. What is the wavelength of these standing waves? Which harmonic has been excited?

21. Two air columns resonate at the same fundamental frequency. Column A is open at both ends and column B has one end open and the other end closed. If the length of column B is 0.4 m, what is the length of column A?

*22. While indoors, a member of the marching band carefully adjusts the length of his flute (an air column with both ends open) to play a frequency of 440 Hz. When he gets outside, he finds that the temperature is 0°C. What is the new frequency of this note?

23. If two sound waves produce a beat frequency of 3 Hz and one of the waves has a frequency of 424 Hz, what are the possible frequencies of the other wave?

24. A guitar string produces 4 beats per second when played with a 250 Hz tuning fork and 9 beats per second when played with a 255 Hz tuning fork. What is the vibrational frequency of the string?

25. A high-pitched whistle has a frequency of 800 Hz. What is the frequency heard by an observer at rest when the whistle is moving away from the observer at 30 m/s?

26. A stationary, high-pitched whistle has a frequency of 800 Hz. What is the frequency heard by an observer moving away from the whistle at 30 m/s?

27. A commuter is standing on the train platform as the express train passes through at 40 m/s. If the train's whistle has a frequency of 400 Hz, what change in frequency does the commuter hear?

28. A passenger in a high-speed train hears a crossing bell with a frequency of 400 Hz (as heard by someone on the ground). If the train is traveling at 40 m/s, what is the shift in frequency heard by the passenger as the train passes the bell?

*29. A police car going 40 m/s is approaching you in a car traveling at 25 m/s. If the siren has a frequency of 400 Hz, what are the frequencies you hear while the police car approaches and after it has passed you?

*30. What frequencies would you hear if the police car in the previous problem overtook you from behind?

*31. As you drive past a stationary speaker, you hear a shift of 50 Hz. If the sound has a frequency of 440 Hz, how fast are you driving?

*32. A person walks between two identical speakers emitting sound waves of 440 Hz. If the person hears a beat frequency of 2 Hz, how fast is the person walking?

17 — LIGHT

17.1 Pinhole Cameras

The size of the image produced by a pinhole camera depends on the object's size and the relative distances of the object and the image from the pinhole. In Figure 17.1.1 two rays are drawn from the top and bottom of the object to the film plane. Because the two rays cross at the pinhole, we know that the angles formed at the crossing are equal, giving us two triangles that have the same shape. Such similar triangles have the property that the ratios of corresponding dimensions are equal. For our purposes, we equate the ratio of the heights to the ratio of the bases. The bases are the object size h_o and image size h_i, while the heights are the distance d_i from the pinhole to the image and the distance d_o from the pinhole to the object.
Symbolically, we have

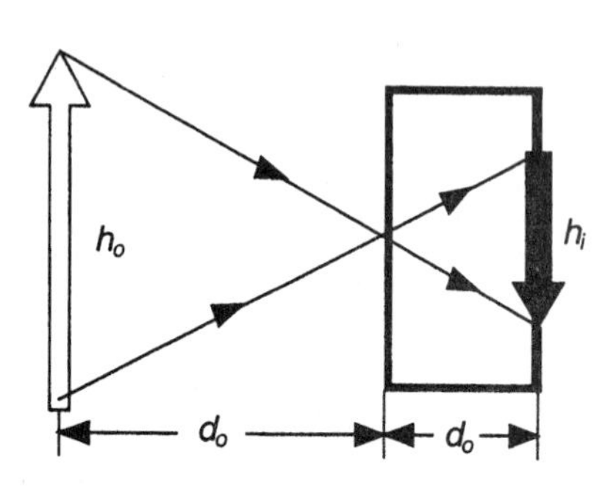

Fig. 17.1.1

$$\frac{h_i}{h_o} = \frac{d_i}{d_o}$$

Example 17.1.1

Calculate the size of the image of a person who is 2 m tall if the person stands 4 m from the pinhole camera. Assume that the back wall of the camera is 0.4 m from the pinhole.

Solving our expression for the image height and plugging in numbers, we obtain

$$h_i = h_o \frac{d_i}{d_o} = 2\text{ m}\left(\frac{0.4\text{ m}}{4\text{ m}}\right) = 0.2\text{ m}$$

Therefore, the image will be 0.2 m tall, which is one-tenth the size of the person.

Practice: What is the image size if the person moves to 2 m from the pinhole?
Answer: 0.4 m

17.2 Flat Mirrors

The image of any object formed by a flat mirror is located behind the mirror on a line that runs from the object perpendicular through the mirror. The image is located a distance behind the mirror equal to the distance the object is in front of the mirror.

Example 17.2.1

Three light bulbs (1, 2, and 3) are placed in front of a mirror. The drawing at the right is a top view; that is, you are looking down on the scene from above. On the diagram, draw in the images of the bulbs at their appropriate locations and indicate the region where an observer could stand in order to see the images of all three light bulbs.

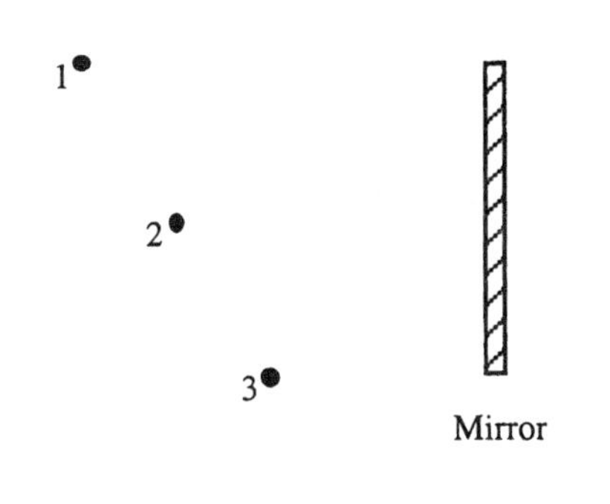

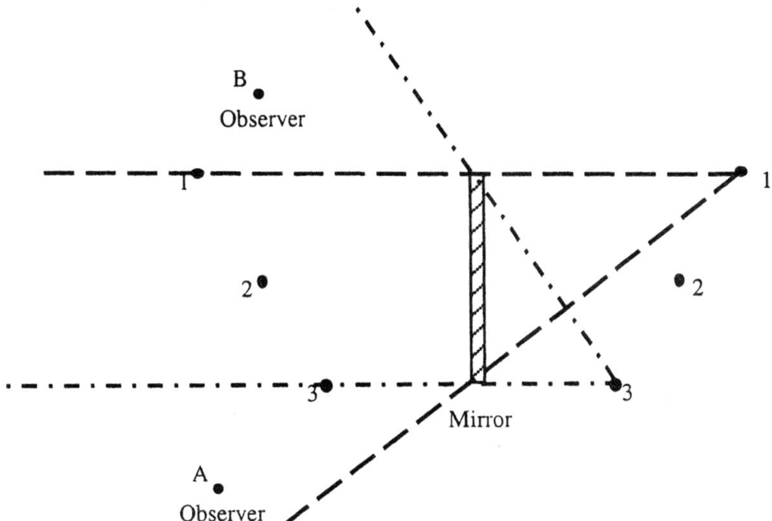

Once the images of the lights have been found, we can think of the mirror as a window that allows us to view the images. The image of light 1 can be seen by an observer anywhere in front of the plane of the mirror between the dashed lines that start at image 1. The observer marked A would be able to see this image. The image of light 3 can be seen by an observer anywhere in front of the plane of the mirror between the dashed lines that start at image 3. The observer marked B would be able to see this image, but not the image of 1. An observer in the overlap region would be able to see all three images.

17.3 Locating the Images

Although ray diagrams are very useful in determining what type of image is produced and roughly where the image is located, an algebraic equation will give us more accurate results. Using geometry and algebra, we can derive an expression for the magnification m and a relationship between the distance s of the object from the surface of the mirror, the distance s' of the image from the surface of the mirror, and the focal length f of the mirror.

In our derivation, we will need to examine various *similar* triangles. These are triangles that have the same shape, but differ in size. Two right triangles are similar if they have one other angle the same. (In effect, this means that all three angles are the same.) Corresponding dimensions of similar triangles are proportional to each other. If the corresponding sides are labeled as in Figure 17.2.1, this can be written

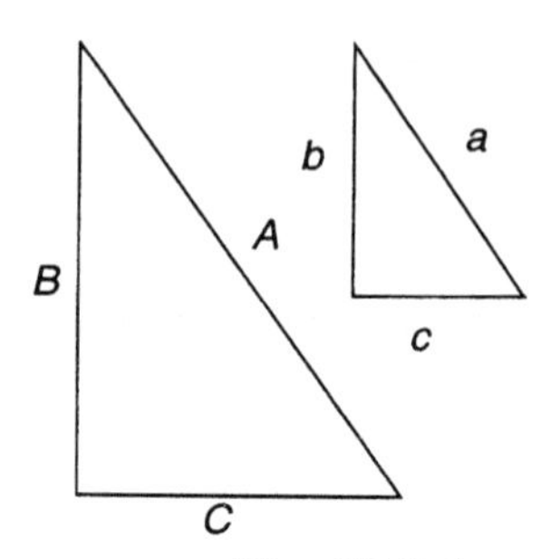

Fig. 17.3.1

$$\frac{a}{A} = \frac{b}{B} = \frac{c}{C}$$

In Figure 17.2.2, we have drawn a ray diagram for locating the image I of an object O formed by a concave mirror with a focal length f. The ray from the top of the object (point H) to the intersection of the optic axis with the surface of the mirror (point M) is not one of our usual rays, but it is useful in the derivation. We designate angles with three letters. For example, consider the ray HM that we just discussed. Its angle of incidence is the angle HMO because the optic axis is normal to the surface. The reflected angle is IMH'. Triangles are designated by the letters at the three corners.

The triangle HOM is similar to triangle $H'IM$ because the angle of reflection IMH' is equal to the angle of incidence HMO. Therefore,

$$\frac{h'}{h} = \frac{s'}{s}$$

where h and h' are the heights of the object and image, respectively. But this ratio is just the magnification m.

$$m \equiv \frac{h'}{h} = \frac{-s'}{s}$$

where the minus sign is inserted to agree with the standard sign convention:

m is positive if the image is erect
m is negative if the image is inverted

To obtain the relationship for the image location, we use triangles HOC and $H'IC$. These triangles are similar because the angles at C are equal to each other because they are formed by crossing straight lines. Notice that the length of the horizontal side of triangle HOC is equal to $s - R$ and the length of the horizontal side of triangle $H'IC$ is $R - s'$. Therefore, these similar triangles give us the following ratios.

$$\frac{h'}{h} = \frac{R - s'}{s - R}$$

We now replace the ratio h'/h with s'/s from our first set of similar triangles.

$$\frac{s'}{s} = \frac{R - s'}{s - R}$$

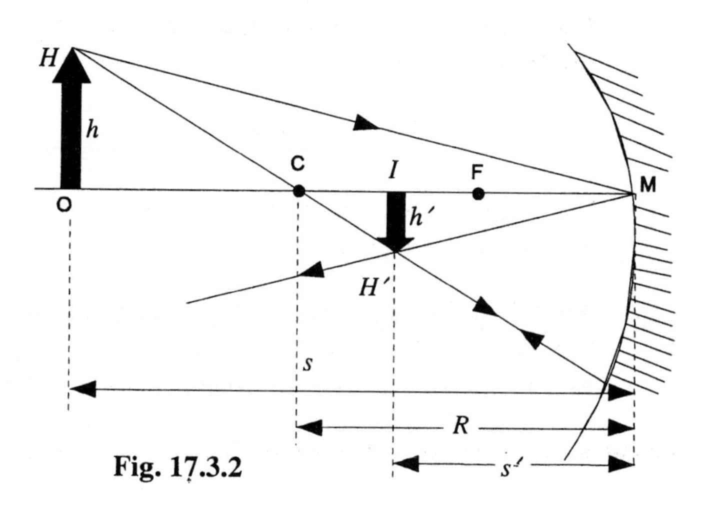

Fig. 17.3.2

We now rearrange the latter expression to get

$$\frac{1}{s}+\frac{1}{s'}=\frac{2}{R}$$

We obtain the mirror formula by recognizing that $f = R/2$.

$$\frac{1}{s}+\frac{1}{s'}=\frac{1}{f}$$

The mirror formula can be used for concave and convex mirrors, provided the sign convention for the focal lengths is used:

f is positive for a concave mirror
f is negative for a convex mirror

If we restrict ourselves to cases with a single mirror and real objects, the sign for the object distance s is always positive. The sign for the image distance s' tells us whether the image is located in front of or behind the mirror and whether it is real or virtual. Virtual images are located behind the mirror, and real images are located in front of the mirror.

s' is positive if the image is in front of the mirror and real
s' is negative if the image is behind the mirror and virtual

Example 17.3.1

The focal length of a concave mirror is 20 cm. An object is located 60 cm in front of the mirror. Where is the image located? What type of image is it? What is the magnification of the image? Is it erect or inverted?

The mirror is concave, so the focal length is positive. Rearranging the mirror formula, we have

$$\frac{1}{s'}=\frac{1}{f}-\frac{1}{s}=\frac{1}{20\text{ cm}}-\frac{1}{60\text{ cm}}=\frac{2}{60\text{ cm}}=\frac{1}{30\text{ cm}}$$

Because s' is positive, the image is real and located in front of the mirror. We can now find the magnification.

$$m=-\frac{s'}{s}=-\frac{30\text{ cm}}{60\text{ cm}}=-\frac{1}{2}$$

The magnification is negative, which means that the image is inverted. The image is one-half as big as the object.

Practice: Describe the image produced when this object is moved to 10 cm from the mirror.
Answer: s' = –20 cm (virtual image behind the mirror) and m = 2 (erect)

Example 17.3.2

The focal length of a convex mirror is 20 cm. An object is located 20 cm in front of the mirror. Describe the image.

The focal length is negative for a convex mirror.

$$\frac{1}{s'} = \frac{1}{f} - \frac{1}{s} = \frac{1}{-20\text{ cm}} - \frac{1}{20\text{ cm}} = \frac{-2}{20\text{ cm}} = \frac{-1}{10\text{ cm}}$$

Because s' is negative, the image is virtual and located behind the mirror. The magnification is given by

$$m = -\frac{s'}{s} = -\frac{-10\text{ cm}}{20\text{ cm}} = \frac{1}{2}$$

The magnification is positive, so we know that the image is erect and one-half as big as the object.

Practice: Describe the image produced when this object is moved to 10 cm from the mirror.
Answer: s' = –6.67 cm (virtual image behind the mirror) and m = 2/3 (erect)

17.4 Speed of Light

Example 17.4.1

How long does it take light from the Sun to reach Earth?

Using the data from the appendix, we have

$$t = \frac{d}{v} = \frac{1.5\times10^{11}\text{ m}}{3\times10^{8}\text{ m/s}} = (500\text{ s})\left[\frac{1\text{ min}}{60\text{ s}}\right] = 8\tfrac{1}{3}\text{ min}$$

Practice: How long does it take light from the Moon to reach Earth?
Answer: 1.28 s

The distances in astronomy are often so large that they are often measured in light-years, the *distance* light travels in one year. Knowing the speed of light, we can calculate the length of a light-year.

$$1\text{ LY} = ct = (3\times10^{8}\text{ m/s})(1\text{ year})\left[\frac{365\text{ days}}{1\text{ year}}\right]\left[\frac{24\text{ h}}{1\text{ day}}\right]\left[\frac{3600\text{ s}}{1\text{ h}}\right] = 9.46\times10^{15}\text{ m}$$

This is a tremendous distance—9.46 trillion km, or almost 6 trillion miles!

Example 17.4.2

The Andromeda galaxy is located about 2.1×10^{19} km from Earth. We are able to see it with our naked eyes because it shines with the light of more than 200 billion stars. What is the distance to the Andromeda galaxy measured in light-years?

$$d = 2.1\times10^{19}\ \text{km}\left[\frac{1\ \text{LY}}{9.46\times10^{12}\ \text{km}}\right] = 2.2\times10^{6}\ \text{LY}$$

Because the Andromeda galaxy is 2.2 million LY away, we know that the light leaving there now will not arrive at Earth for 2.2 million years.

Practice: The brightest star that we observe in the night sky is Sirius. One of the reasons that it appears so bright is that it is relatively close to Earth. If the distance to Sirius is 8.1×10^{13} km, how long does it take light from Sirius to reach us?
Answer: 8.56 years

Problems

1. A landscape painter places a camera obscura 20 m from a 8-m tall tree. How big is the image if the back wall is 2 m away from the pinhole?
2. A camera obscura used by a portrait painter is located 5 m from a child who stands 1 m tall. How big is her image if the back of the camera obscura is 1.5 m from the pinhole?
3. An artist would like to paint a portrait that is one-fourth the size of the person. If the distance to the screen is 0.5 m from the pinhole, how far from the pinhole should the person stand?
4. What distance from the pinhole to the back wall of a camera obscura is needed to produce an image of the full moon that is 10 cm in diameter?

*5. The image in a camera obscura is 12 cm tall when the screen is placed 75 cm from the pinhole. If the artist wishes to produce an image of the same object that is 38 cm tall without moving the object, how far from the pinhole should the screen be placed?

6. The image of a barn in a camera obscura is 0.4 m tall. If the barn is 50 m from the pinhole and the screen is 2 m behind the pinhole, how tall is the barn?
7. A concave mirror has a focal length of 30 cm. What kind of image is produced by an object located 60 cm from the mirror? Where is it located and what is its magnification?
8. A 15-cm ruler is placed 60 cm in front of a concave mirror with a focal length of 40 cm. Where is the image located and how big is it?

*9. You stand with your nose 5 cm from the surface of a concave shaving mirror. The virtual image of your nose appears magnified 3 times. What is the mirror's focal length?

10. A 8-cm tall candle is placed 20 cm from a concave mirror that has a focal length of

60 cm. How big is the image and where is it located?

*11. A 12-cm candle is placed 60 cm in front of a concave mirror with focal length 15 cm. Where is the image located and how big is it? Would it be upright or inverted?

*12. You wish to use a concave mirror of focal length 20 cm to produce a 15-cm tall image located 25 cm from the mirror. Where should you place the object and how tall should it be?

*13. Where is the image of a candle placed at the focal point of a concave mirror?

*14. Where in front of a concave mirror would you place a candle so that it appears to burn at both ends?

*15. Your thumb's image, when viewed in a convex mirror, appears one-third its normal size. If your thumb is 15 cm from the surface of the mirror, what is the mirror's focal length?

16. If your nose is 12 cm from the surface of a shiny metal sphere with a diameter of 8 cm, where is the image of your nose?

*17. Show that a convex mirror cannot produce a real image of an object.

*18. Where would you place an object in front of a convex mirror so that its image is one-half the size of the object?

19. The star Aldebaran, the red eye of Taurus, the Bull, is 4.9×10^{14} km from Earth. How long does it take its light to reach Earth?

20. If all of the stars were somehow placed at the same distance from the Sun, the star Rigel in Orion would be the brightest star in the heavens. It appears dimmer than other stars because it is about 7.7×10^{15} km away. If Rigel suddenly blew apart, how long would it be before we would know about it?

21. Supernova 1987A occurred at a distance of 1.6×10^{18} km from Earth. How long ago did it occur?

22. If the diameter of the Milky Way Galaxy is 120,000 LY, how far is it across in km?

18 — REFRACTION OF LIGHT

18.1 Index of Refraction

If you measure the depth of the image produced in the water in Figure 18-4(c) in the text, you will discover that the image depth is three-fourths the depth of the object. It is interesting to note that the index of refraction of water is four-thirds; that is, it is the reciprocal of the relative depth. Similar drawings for other substances such as glass verify that the depth d' of the image is given by

$$d' = \frac{d}{n}$$

where d is the depth of the object and n is the index of refraction.

Example 18.1.1

A decal is pasted on the far side of a thick piece of glass. If the glass is 6 cm thick, how far into the glass does the image of the decal appear?

The index of refraction for the glass that we used in the text is 1.5; therefore

$$d' = \frac{d}{n} = \frac{6\text{ cm}}{1.5} = 4\text{ cm}$$

Practice: Where is the image if the glass has an index of refraction equal to 2?
Answer: 3 cm

Example 18.1.2

A fish looks straight up and sees the tip of a spear that appears to be 20 cm above the surface of the water. Where is the tip of the spear?

A simple ray diagram convinces us that the tip of the spear must be closer to the water than it appears. Again, the index of refraction relates the image location to the object location.

$$d = \frac{d'}{n}$$

Because the index of refraction for water is four-thirds, the tip of the spear is actually closer by a factor of three-fourths; it is located 15 cm above the surface of the water.

18.2 Images Produced by Lenses

Although ray diagrams are very useful in determining what type of image is produced and roughly where the image is located, algebraic equations give us more accurate results. Using geometry and algebra, we can show that the formulas we used for mirrors in Section 17.2 can also be used for lenses with minor changes in definition. We let s be the distance from the object to the center of the lens, s' be the distance of the image from the center of the lens, and f be the focal length of the lens as shown in Figure 18.2.1.

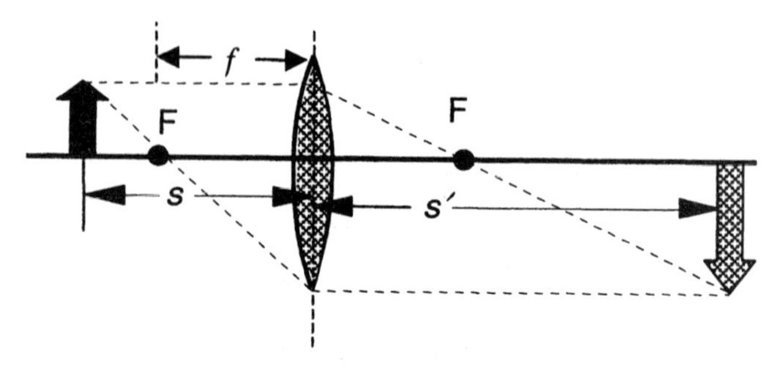

Fig. 18.2.1

The lens formula is identical to the mirror formula

$$\frac{1}{s}+\frac{1}{s'}=\frac{1}{f}$$

with following conventions for the focal length:

f is positive for a converging lens
f is negative for a diverging lens

Once again the sign of the image distance tells us the location of the image and whether it is real or virtual. If the object is located on the near side of the lens, images on the far side are real and images on the near side are virtual:

s' is positive if the image is on the far side of the lens and real
s' is negative if the image is on the near side of the lens and virtual

We can also use the same equation for the magnification of the image

$$m=-\frac{s'}{s}$$

with the sign convention that:

m is positive if the image is erect
m is negative if the image is inverted

Example 18.2.1

The focal length of a converging lens is 20 cm. An object is located 60 cm from the lens. Where is the image located? What type of image is it? What is the magnification of the image? Is it erect or inverted?

Because the lens is converging, the focal length is positive. Rearranging the lens formula, we have

$$\frac{1}{s'} = \frac{1}{f} - \frac{1}{s} = \frac{1}{20\text{ cm}} - \frac{1}{60\text{ cm}} = \frac{3-1}{60\text{ cm}} = \frac{2}{60\text{ cm}} = \frac{1}{30\text{ cm}}$$

$$s' = 30\text{ cm}$$

s' is positive, indicating that the image is real. It is located 30 cm from the lens on the far side (opposite the object). The magnification is

$$m = -\frac{s'}{s} = -\frac{30\text{ cm}}{60\text{ cm}} = -\frac{1}{2}$$

We know that the image is inverted because the magnification is negative. The image is one-half as big as the object.

Practice: Describe the image produced when this object is moved to 10 cm from the lens.
Answer: s' = –20 cm (virtual image on the near side) and m = 2 (erect)

Example 18.2.2

The focal length of a diverging lens is 20 cm. An object is located 20 cm from the lens. Locate and describe the image.

The focal length is negative for a diverging lens.

$$\frac{1}{s'} = \frac{1}{f} - \frac{1}{s} = \frac{1}{-20\text{ cm}} - \frac{1}{20\text{ cm}} = \frac{-2}{20\text{ cm}} = \frac{-1}{10\text{ cm}}$$

$$s' = -10\text{ cm}$$

Therefore, the image is virtual and located 10 cm from the lens on the near side with a magnification of

$$m = -\frac{s'}{s} = -\frac{-10\text{ cm}}{20\text{ cm}} = \frac{1}{2}$$

The positive magnification tells us that the image is erect.

Practice: Describe the image produced when this object is moved to 10 cm from the lens.
Answer: s' = –6.67 cm (virtual image on the near side of the lens) and m = 0.667 (erect)

18.3 Cameras

Example 18.3.1

A camera with a 50-mm lens is used to take a portrait of a basketball player with a height of 2 m. If the player stands 2 m from the lens, where is the image formed and how large will it be?

Using the results from the previous section and setting $f = 5$ cm and $s = 200$ cm, we calculate the image location s' to be

$$\frac{1}{s'} = \frac{1}{f} - \frac{1}{s} = \frac{1}{5\text{ cm}} - \frac{1}{200\text{ cm}} = \frac{39}{200\text{ cm}}$$

$$s' = \frac{200\text{ cm}}{39} = 5.13\text{ cm} = 51.3\text{ mm}$$

Therefore, the lens should be adjusted to be 51.3 mm from the film. Notice that this is very close to the focal length of the lens. As the subject moves farther away, the term $1/s$ approaches zero and the image distance approaches the focal length. Therefore, for objects at infinity, the film should be located at the focal point.

We can calculate the image size using the magnification.

$$h' = h\frac{(-s')}{s} = (2\text{ m})\frac{(-51.3\text{ mm})}{2\text{ m}} = -51.3\text{ mm}$$

Because the film is only 35-mm wide, you cannot take a full view without turning the camera on its side. However, if you switch to a wide angle lens with a 28-mm focal length, the image will fit.

Practice: What is the image size using the 28-mm lens?
Answer: 28.4 mm high

18.4 Our Eyes

Example 18.4.1

What is the focal length of a lens rated at +64 diopters?

$$f = \frac{1\text{ m}}{d} = \frac{1\text{ m}}{64} = 0.0156\text{ m} = 1.56\text{ cm}$$

Example 18.4.2

If your eyeball is 1.67 cm long, how close must an object be to form a focused image on the retina when the lens in the eye has the maximum 64 diopters?

Setting $s' = 1.67$ cm and using the focal length from the previous example, we have

$$\frac{1}{s} = \frac{1}{f} - \frac{1}{s'} = \frac{1}{1.56\text{ cm}} - \frac{1}{1.67\text{ cm}} = 0.0422/\text{cm}$$

$$\Rightarrow s = 23.7\text{ cm}$$

18.5 Magnifiers

Your eyes are not able to focus on objects closer than the *near point* of your eyes. This puts a limit on the angular size of the object. If you bring it closer to increase the size of its image on your retina, the image becomes blurred. However, if you place a converging lens so that the object is at its focal point, the lens produces an image at infinity that is easy to look at. Moreover, the image on your retina is bigger. If the near point for your eyes is d_n, the magnification of the simple magnifier is given by

$$m = \frac{d_n}{f}$$

where f is the focal length of the lens.

Example 18.5.1

If your near point is 24 cm and the magnifying glass has a focal length of 6 cm, what is the magnification?

$$m = \frac{d_n}{f} = \frac{24\text{ cm}}{6\text{ cm}} = 4$$

This is the largest practical magnification that you can get with a single lens.

Problems

1. Where is the image of a decal on the bottom of a 3-m deep swimming pool?
2. A restaurant has a dividing wall made of 15-cm thick glass bricks. Your kid brother pushes his nose up against the other side of the wall and you estimate that the image of the end of his nose is only 9 cm from your side of the wall. Estimate the index of refraction of this glass.
3. A swimmer lying on the bottom of a swimming pool observes a light on the ceiling. If

the ceiling is 3 m tall, how far above the water does the light appear to be?

4. A fish is swimming behind a thick glass window ($n = 1.5$) used as a viewing port into an aquarium. If the fish is next to the window and the window is 2 cm thick, how far behind the front of the window does the fish appear to be?

5. A 15-cm ruler is placed 75 cm from a converging lens with a focal length of 60 cm. Where is the image located and how big is it?

6. A converging lens has a focal length of 30 cm. What kind of image is produced by an object located 60 cm from the lens? Where is the image located and what is its magnification?

7. Where is the image located if an object is placed 15 cm away from a converging lens with a focal length of 30 cm?

8. You look at your thumb through a converging lens and its virtual image appears to be three times the height of your thumb. If your thumb is 10 cm from the lens, what is the focal length of the lens?

9. A converging lens with a focal length of 10 cm is used to make a slide projector. If the image is to be projected onto a screen 5 m away, how far from the lens should the slide be placed?

10. An overhead projector is made from a converging lens with a focal length of 20 cm. If the image is projected onto a screen at a distance of 3 m, how far should the lens be from the transparency?

11. If your nose is 10 cm from a diverging lens with a focal length of 5 cm, where is the image of your nose?

12. A diverging lens with a focal length of 45 cm is placed 60 cm from a 4-cm tall candle. Where is the image of the candle and how big is it?

13. A diverging lens has a focal length of 30 cm. What kind of image is produced by an object located 20 cm from the lens? Where is the image located and what is its magnification?

14. Where is the image located if an object is placed 10 cm away from a diverging lens with a focal length of 30 cm? What is the magnification?

15. The virtual image of a candle appears 10 cm behind a diverging lens of focal length 30 cm. The image appears to be 20 cm tall. Where is the candle located and how tall is it?

16. A 6-cm chess piece appears only 2 cm tall when viewed through a diverging lens. If the chess piece is located 45 cm from the lens, what is the focal length of the lens?

17. A tourist in Yellowstone National Park takes a photograph of a buffalo that is 40 m away and 2.2 m tall. If the lens has a focal length of 50 mm, how large is the image of the buffalo on the film?

18. If the tourist in the previous problem switches to a telephoto lens with a focal length of 200 mm, how large will the image of the buffalo be?

19. What is the size of the image on the retina if a person is looking at a cowboy who is 3 m away and stands 180 cm tall?

20. A person is looking at a house that is 12 m tall to the top of the roof. If the person is 50 m away, how tall is the image on the retina?

21. The retina in a child's eye is 1.67 cm from the lens. What is the strength of the child's lens (in diopters) when forming a clear image of a distant mountain range? *Hint*: Treat the distance to the object as infinite.

22. If the nearest object that the child in the previous problem can focus clearly is 8 cm from her lens, what is the maximum strength of her eyes (in diopters)?

23. If a converging lens (+4 diopters) and a diverging lens (–6 diopters) are glued together, what is the focal length of the pair?
24. A converging lens rated at +8 diopters is bonded to a diverging lens of –4 diopters. What is the focal length of the combination?
25. A 10-year-old child can focus as close as 7 cm. What focal length lens should the child use to get a magnification of 3?
26. A senior citizen has a near point that is 200 cm. What focal length lens should the senior citizen use to get a magnification of 3?

19 — A MODEL FOR LIGHT

19.1 Refraction

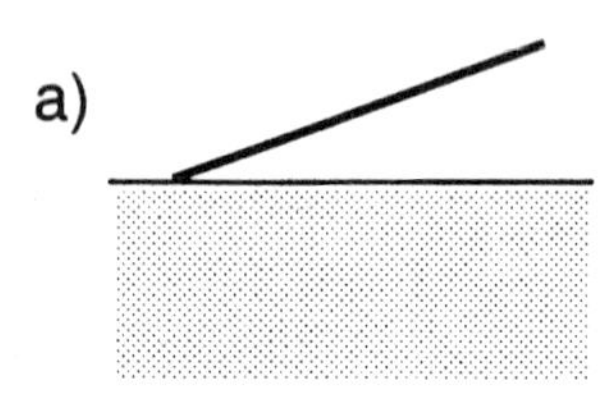

As discussed in the text, the speed of light in materials is slower than the speed of light in a vacuum. Michelson showed that the speed v in a material is given by

$$v = \frac{c}{n}$$

where c is the speed in a vacuum and n is the index of refraction of the material. It is this change in speed that is responsible for the change in direction that takes place at the surface. Imagine a single wave front hitting the surface at an angle, as shown in Figure 19.1.1. The wave is moving at right angles to the front. Let the shaded region have the higher index of refraction, and therefore, the slower speed.

In (a), one end of the wave front has just reached the surface. From now on, this portion of the wave moves slower. At some time later (part b), the wave front exhibits a bend at the surface. (We've indicated the previous position of the wave front with a dashed line.) In (c), the entire wave front is traveling in the new direction.

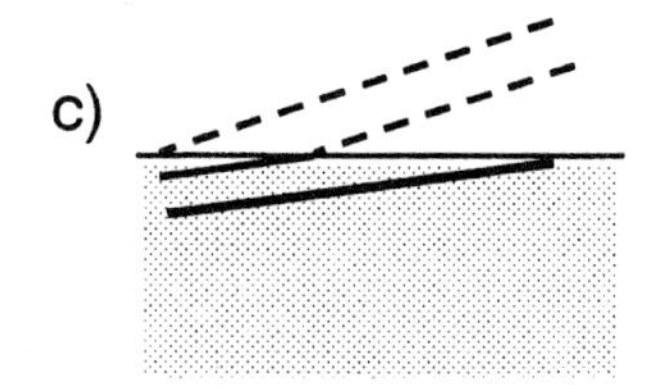

Fig. 19.1.1

You should be able to convince yourself that if you read these diagrams in the opposite order, they correspond to a wave front leaving a material with a higher index of refraction and bending away from the normal.

Example 19.1.1

What is the speed of light in a diamond with an index of refraction of 2.42?

$$v = \frac{c}{n} = \frac{3\times10^8 \text{ m/s}}{2.42} = 1.24\times10^8 \text{ m/s}$$

Practice: What is the speed of light in zircon (fake diamond) with an index of 1.92?
Answer: 1.56×10^8 m/s

Because the frequency of the light does not change as it crosses the boundary between two transparent materials, the change in speed requires a corresponding change in the wavelength. The wavelength must get shorter as the light slows down because the crests cannot travel as far during one period. The wavelength λ_n in a material with an index of refraction n is given by

$$\lambda_n = \frac{\lambda}{n} = \frac{c}{nf}$$

where λ is the wavelength in a vacuum.

Example 19.1.2

What is the wavelength of red light (f = 4.5 × 10[14] Hz) in diamond (n = 2.42)?

$$\lambda_n = \frac{c}{nf} = \frac{3\times10^8 \text{ m/s}}{(2.42)(4.5\times10^{14} \text{ Hz})} = 275 \text{ nm}$$

The wavelength of this light is 667 nm in a vacuum.

Practice: What is the wavelength of blue light (f = 7 × 10[14] Hz) in diamond?
Answer: 177 nm

19.2 Interference

The equation that we used for the interference of water waves in Section 14.8 holds for all types of waves. Therefore, we can use the equation with light.

$$y_{max} = m\frac{\lambda L}{d} \qquad m = 0, \pm1, \pm2, \pm3, \ldots$$

where y_{max} is the distance measured along the wall from the central bright region to another bright (antinodal) region, λ is the wavelength, L is the distance to the wall, d is the spacing of the slits, and m is an integer that numbers the antinodal regions.

Example 19.2.1

The light from a helium-neon laser has a wavelength of 633 nm. If the laser beam shines on two narrow slits separated by 0.1 mm, what is the spacing of the interference antinodes on a wall 10 m away?

Because we know that $y_{max} = 0$ for $m = 0$, we can calculate the location of the first maximum ($m = 1$) to find the spacing. Being careful to express the lengths in meters (or at least units that will cancel), we have

$$y_{max} = m\frac{\lambda L}{d} = (1)\frac{(6.33\times10^{-7} \text{ m})(10 \text{ m})}{1\times10^{4} \text{ m}} = 6.33\times10^{-2} \text{ m}$$

Therefore, the bright regions will be 6.33 cm apart on the wall.

Practice: What is the spacing on the wall if light from a sodium vapor lamp (λ = 589 nm) is used?
Answer: 5.89 cm apart

Example 19.2.2

The spacing of the bright lines in an interference pattern is measured to be 2.2 cm. If the wall is 1.5 m away from slits with a spacing of 0.05 mm, what is the wavelength of the light?

$$\lambda = \frac{y_{max}\, d}{mL} = \frac{(2.2\times10^{-2}\text{ m})(5\times10^{-5}\text{ m})}{(1)(1.5\text{ m})} = 733\text{ nm}$$

19.3 Diffraction

We can also use our expression for the diffraction pattern from Section 15.8 to analyze the diffraction of light.

$$y_{min} = m\frac{\lambda L}{w} \qquad m = \pm1, \pm2, \pm3, \ldots$$

where y_{min} is the distance measured along the wall from the central bright region to the nodal regions, λ is the wavelength, L is the distance to the wall, w is the width of the slit, and m is an integer that numbers the nodal regions. Notice that $m = 0$ is excluded. This occurs because the central region is an antinodal region. This central region is twice as wide as the other antinodal regions and is much brighter than those on either side.

Example 19.3.1

How wide is the central maximum produced by yellow light with a wavelength of 600 nm striking a slit with a width of 0.09 mm if the slit is 180 cm from the viewing screen?

We know that the minima on either side of the central maximum are given by our formula. Therefore, the full width of the maximum is given by the difference in y_{min} with $m = 1$ and $m = -1$. For $m = 1$, we have

$$y_{min} = m\frac{\lambda L}{w} = (1)\frac{(6\times10^{-7}\text{ m})(180\text{cm})}{0.009\text{ cm}} = 1.2\times10^{-2}\text{ m}$$

Therefore, one minimum is at 1.2 cm. The minimum on the other side must be at −1.2 cm. This gives a full width of 2.4 cm for the central maximum.

Practice: What happens to the width of the central maximum if the screen is moved twice as far away?
Answer: It doubles to 4.8 cm.

19.4 Thin Films

When light shines on a thin film in air, the waves that reflect from the front surface are inverted, while those reflecting from the back surface are not inverted (Figure 19-11 in the text). This means that a very thin film (much thinner than the wavelength of the light) will not reflect light because the two reflected waves will interfere destructively.

If the thickness of the film is increased to equal one-fourth of the wavelength of the light measured *in the film*, the waves reflecting off the back surface travel an extra one-half wavelength and emerge in phase with those reflected from the front surface. This means that the light will be strongly reflected. Every time the thickness of the film is increased by an additional half wavelength, the wave inside the film travels an extra wavelength and continues to emerge in phase. The following thicknesses will yield strong reflections.

$$d_m = \frac{\lambda_f}{4} + m\frac{\lambda_f}{2} = \left(m + \tfrac{1}{2}\right)\frac{\lambda_f}{2} \qquad m = 0, 1, 2, 3, \ldots$$

where λ_f is the wavelength in the film

$$\lambda_f = \frac{\lambda}{n}$$

with λ the wavelength in a vacuum and n the index of refraction of the film.

If a film is coated onto a material with a higher index of refraction, the waves are inverted at the back surface. In this case, a film with "zero" thickness would reflect light strongly. Increases in the thickness of one-half wavelength would also produce strong reflection. Therefore, the possible thicknesses for strong reflections are

$$d_m = m\frac{\lambda_f}{2} \qquad m = 1, 2, 3, \ldots$$

Example 19.4.1

What thicknesses of a soap film will strongly reflect light from a helium-neon laser with a wavelength of 633 nm in air?

If we assume that the index of a soap film is close to that of water, the wavelength of the light in the film is

$$\lambda_f = \frac{\lambda}{n} = \frac{633\text{ nm}}{1.33} = 476\text{ nm}$$

Because the soap film is in air, we use our first equation to calculate the thicknesses.

$$d_m = \left(m + \tfrac{1}{2}\right)\frac{\lambda_f}{2} = \left(m + \tfrac{1}{2}\right)\frac{476\text{ nm}}{2} = \left(m + \tfrac{1}{2}\right)(238\text{ nm})$$

$$d_0 = \tfrac{1}{2}(238\text{ nm}) = 119\text{ nm}$$

$$d_1 = \tfrac{3}{2}(238\text{ nm}) = 357\text{ nm}$$

$$d_2 = \tfrac{5}{2}(238\text{ nm}) = 595\text{ nm} \quad \text{and so on.}$$

Practice: What is the thinnest soap film that will strongly reflect green light with a wavelength of 550 nm?
Answer: 103 nm

Example 19.4.2

A camera lens is coated with a thin film of magnesium fluoride with an index of refraction of 1.38. What thickness should be used to reflect violet light with a wavelength of 414 nm?

Because the film has an index of refraction that is intermediate between the air and the glass, inversions will take place at both surfaces of the film. Therefore, we use our second equation to obtain the desired thickness.

$$d_m = m\frac{\lambda_f}{2} = (1)\frac{\lambda}{2n} = \frac{414\text{ nm}}{2(1.38)} = 150\text{ nm}$$

Practice: What is the next thickness that will strongly reflect this light?
Answer: 300 nm

Problems

1. What is the speed of light in carbon tetrachloride, which has an index of refraction of 1.46?
2. If it takes light 2 ns to traverse 40 cm of glass, what is the index of refraction of the glass?
3. Upon entering material X, light slows down by 30% compared to its speed in vacuum. What is the index of refraction of material X?
4. By what fraction does light slow down in passing from air into glass ($n = 1.6$)?

*5. If light ($\lambda = 633$ nm) travels through 1 mm of glass ($n = 1.5$), by how many wavelengths will it be delayed compared to a vacuum?

6. By how much is light delayed in traveling through 2 m of glass ($n = 1.6$) compared to a vacuum?
7. What is the wavelength of the green light from a mercury discharge tube ($f = 5.49 \times 10^{14}$ Hz) in diamond?
8. Red light from a hydrogen discharge tube has a frequency of 4.57×10^{14} Hz. What is the wavelength of this light in zircon ($n = 1.92$)?
9. In a two-slit interference experiment using an argon laser ($\lambda = 515$ nm), the first maximum is located 4 cm from the centerline when the screen is placed 10 m from the slits. At the location of this first maximum, how much farther away is the far slit than the near slit?
10. For the previous problem, find the slit separation.
11. Green light from a mercury discharge tube ($\lambda = 546$ nm) shines on two slits separated by 0.2 mm. What is the location of the second maximum on a screen 4 m away?

12. Blue-green light from an argon laser (λ = 515 nm) produces an interference pattern on a screen located 1.2 m from two slits with a separation of 0.1 mm. What is the spacing of the pattern?

13. The red light from a hydrogen discharge tube produces an interference pattern in which the third maximum is 5.5 cm from the central maximum. If the screen is 0.75 m away and the wavelength is 656 nm, what is the spacing of the slits?

14. In a double slit interference experiment, the third maximum is located 4.8 cm from the central maximum. If the screen is located 2 m from the slits and the slits have a separation of 0.07 mm, what is the wavelength of the light?

15. A 0.2-mm slit is illuminated with yellow light (λ = 588 nm) from a helium discharge tube. If the screen is 0.5 m away, what is the width of the central maximum?

16. What is the width of the central maximum of the diffraction pattern produced by light with a wavelength of 600 nm shining on a slit (w = 0.05 mm) if the viewing screen is 1.4 m away?

17. The spacing of the bright lines in a single slit diffraction pattern (excluding the central bright spot) is 2.6 mm. If the screen in located 3 m from the slit and the slit width is 0.6 mm, what is the wavelength of the light source?

18. The red light from a hydrogen discharge tube with a wavelength of 656 nm produces a diffraction pattern with a central maximum that is 1.5 cm wide. If the screen is 0.5 m away, what is the width of the slit?

19. What is the thinnest soap film that will strongly reflect yellow light with a wavelength of 589 nm?

20. If the light reflected from the two surfaces of a thin film is exactly out of phase, the reflection will be weak and the transmission will be strong. What is the thinnest soap film (other than zero) that will strongly transmit light with a wavelength of 589 nm?

21. What are the possible thicknesses of a benzene layer on water that will strongly reflect green light of wavelength 520 nm in air? The index of refraction for benzene is 1.5.

22. Repeat the previous problem for a benzene layer on flint glass (n = 1.7).

23. A camera lens is coated with a thin film of magnesium fluoride (index = 1.38) to minimize the reflection of yellow light with a wavelength of 600 nm in air. How thick is the film?

*24. A camera lens is coated with a 500-nm film of magnesium fluoride with an index of refraction of 1.38. Which wavelengths in the visible range have minimal reflections?

20 — ELECTRICITY

20.1 The Electric Force

Calculating an electric force is just a matter of plugging the given values into Coulomb's law. However, the ratio of two electric forces can often be obtained without knowing the values of the variables, providing you know the relative values. Let's write down Coulomb's law for the original value of the force using the subscript o on each variable. Then the original force F_o is given by

$$F_o = k\frac{Q_o q_o}{r_o^2}$$

where the two charges Q_o and q_o are separated by a distance r_o.

If we write down the same expression for the new force in terms of the new charges and the new distance of separation using the subscript n, we have

$$F_n = k\frac{Q_n q_n}{r_n^2}$$

Taking the ratio of the two equations and canceling the common factor k, we obtain

$$\frac{F_n}{F_o} = \frac{Q_n q_n r_o^2}{Q_o q_o r_n^2}$$

Whenever a quantity stays the same, it can be canceled in the ratio on the right-hand side. If we then express the other new values in terms of the old values, we can obtain the ratio of the two forces.

Example 20.1.1

What happens to the force if the distance between two charged objects is tripled?

Assuming that the charges stay the same, that is, $Q_n = Q_o$ and $q_n = q_o$, we can cancel them. Tripling the distance is the same as setting $r_n = 3r_o$. Therefore,

$$\frac{F_n}{F_o} = \frac{Q_n q_n r_o^2}{Q_o q_o r_n^2} = \frac{r_o^2}{r_n^2} = \frac{r_o^2}{(3r_o)^2} = \frac{1}{9}$$

The new force has one-ninth the value of the old force. This makes sense because tripling the distance should cut the force by $3^2 = 9$.

Example 20.1.2

What happens to the force if the size of each charge is doubled and their signs are reversed?

This means that $Q_n = -2Q_o$ and $q_n = -2q_o$. Therefore,

$$\frac{F_n}{F_o} = \frac{Q_n q_n r_o^2}{Q_o q_o r_n^2} = \frac{Q_n q_n}{Q_o q_o} = \frac{(-2Q_o)(-2q_o)}{Q_o q_o} = 4$$

Notice that the sign changes don't change the direction of the force.

Practice: What happens to the force if the charges and the separation are all doubled?
Answer: The force stays the same.

20.2 Electricity and Gravity

In the text we calculated the gravitational and electric forces between an electron and a proton separated by a specified distance. Then we took the ratio of the two forces to show that the electric force was approximately 10^{39} times larger. An alternate way of calculating the ratio is to use the technique from the previous section. Writing the ratio symbolically, we have

$$\frac{F_e}{F_g} = \frac{kQ_1Q_2}{GM_1M_2}$$

where the distance of separation r cancels because it is the same for both forces.

Example 20.2.1

What is the ratio of the electric force between an electron and a proton to the gravitational force between them?

$$\frac{F_e}{F_g} = \frac{kQ_1Q_2}{GM_1M_2}$$

$$\frac{F_e}{F_g} = \frac{\left(8.99\times10^9 \frac{\text{N}\cdot\text{m}^2}{\text{C}^2}\right)\left(1.6\times10^{-19}\text{ C}\right)^2}{\left(6.67\times10^{-11}\frac{\text{N}\cdot\text{m}^2}{\text{kg}^2}\right)\left(9.11\times10^{-31}\text{ kg}\right)\left(1.67\times10^{-27}\text{ kg}\right)} = 2.27\times10^{39}$$

which agrees with the answer in the text.

20.3 The Electric Field

By definition, the electric field at a point in space is equal to the force on a unit positive charge placed at that point. In simple cases it can be obtained by calculating the force on a charge q and then dividing by q.

$$E = \frac{F}{q} = \frac{1}{q} k \frac{Qq}{r^2} = k \frac{Q}{r^2}$$

We see that the value of the electric field is independent of the value of the test charge q.

Example 20.3.1

What is the electric field at a distance of 10 cm from a point charge of 4 μC?

$$E = k \frac{Q}{r^2} = \left(8.99 \times 10^9 \frac{\text{N} \cdot \text{m}^2}{\text{C}^2}\right) \frac{4 \times 10^{-6} \text{ C}}{(0.10 \text{ m})^2} = 3.6 \times 10^6 \frac{\text{N}}{\text{C}}$$

Because the electric field is a vector, we must give its direction to completely specify it. The field points directly away from a positive charge because that is the direction of the force on a positive charge placed at the field point.

If a charge of –5 μC is placed at this point, what force will it experience?

Because $E = F/q$, we have $F = qE$. Therefore,

$$F = qE = \left(-5 \times 10^{-6} \text{ C}\right)\left(3.6 \times 10^6 \frac{\text{N}}{\text{C}}\right) = -18 \text{ N}$$

The minus sign means that the force is in the direction opposite the field, that is, directed toward the original charge. This makes sense because the two charges have opposite signs and they will attract each other.

Practice: What is the electric field 20 cm from a point charge of –20 μC?
Answer: 4.5 × 10^6 N/C toward the charge

When two or more charges produce an electric field, we use the principle of superposition; we find the electric field due to each charge and add the contributions as vectors.

Example 20.3.2

Two fixed charges, $Q_A = +4 \times 10^{-8}$ C and $Q_B = -2 \times 10^{-8}$ C, are located as shown below with d_1 = 0.05 m and d_2 = 0.1 m.

Find the electric field at point x.

First, pretend that Q_A is the only charge creating the field. Its contribution is

$$E_A = k\frac{Q_A}{d_1^2} = \left(8.99\times10^9\ \text{N}\cdot\text{m}^2/\text{C}^2\right)\frac{4\times10^{-8}\ \text{C}}{(0.05\ \text{m})^2} = 1.44\times10^5\ \text{N/C}$$

This contribution is to the left because the electric field always points away from a positive charge. Now pretend that Q_B is the only charge creating the field. Its contribution is

$$E_B = k\frac{Q_B}{(d_1+d_2)^2} = \left(8.99\times10^9\ \text{N}\cdot\text{m}^2/\text{C}^2\right)\frac{2\times10^{-8}\ \text{C}}{(0.15\ \text{m})^2} = 7.99\times10^3\ \text{N/C}$$

Note that we used the magnitude of Q_B (without the minus sign) because we only needed the magnitude of E_B. We figure out its direction separately. The electric field points toward negative charges, so this contribution points to the right. The net electric field at point X is the vector sum of these two contributions.

$$E_{net} = E_A + E_B = -1.44\times10^5\ \text{N/C} + 7.99\times10^3\ \text{N/C} = -1.36\times10^5\ \text{N/C}$$

where we have taken vectors pointing to the right as positive. Therefore, the electric field at point X points to the left.

Practice: What force would a charge $q = -5.0\times10^{-9}$ C experience if placed at x?
Answer: 6.8 x 10^{-4} N (to the right)

Example 20.3.3

Two charges, $Q_A = -6\times10^{-9}$ C and Q_B unknown, are located d_2 = 20 cm apart as shown below. When a third charge $Q_C = 4\times10^{-9}$C is placed at point x, at distance d_1 = 10 cm from Q_B, the electric force on Q_C is zero. What is the charge Q_B?

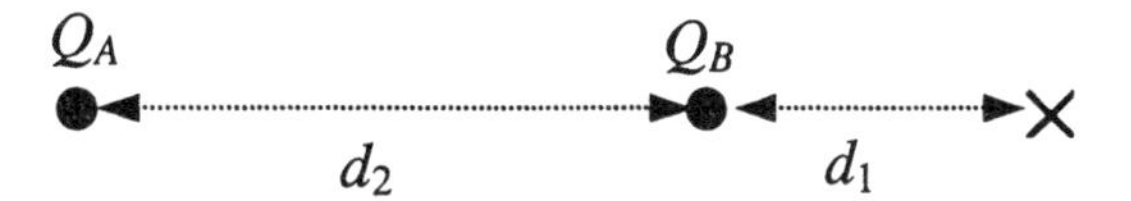

Because $F = qE$, the electric field at X must be zero. This means that the electric field contributions due to Q_A and Q_B must cancel at X. The electric field due to Q_A is

$$E_A = k\frac{Q_A}{(d_1+d_2)^2} = \left(8.99\times10^9\ \frac{\text{N}\cdot\text{m}^2}{\text{C}^2}\right)\frac{6\times10^{-9}\ \text{C}}{(0.3\ \text{m})^2} = 599\ \text{N/C}$$

and points to the left because Q_A is negative. The electric field due to Q_B must have the same magnitude and point in the opposite direction. Therefore, Q_B must be a positive charge with magnitude

$$Q_B = \frac{E_B d_1^2}{k} = \frac{(599\ \text{N/C})(0.1\ \text{m})^2}{\left(8.99\times10^9\ \text{N}\cdot\text{m}^2/\text{C}^2\right)} = 6.66\times10^{-10}\ \text{C}$$

20.4 Electric Potential

Example 20.4.1

If it requires 10 J of work to take an object with a charge of 2 C between points A and B, what is the electric potential difference between A and B?

The electric potential difference is equal to the work required divided by the charge on the object.

$$V = \frac{W}{q} = \frac{10\text{ J}}{2\text{ C}} = 5\frac{\text{J}}{\text{C}} = 5\text{ V}$$

Practice: What would the potential difference be if it required 10 J of work to take a charge of –2 C between A and B?
Answer: –5 V

When parallel conducting plates are connected to the terminals of a battery, the electric potential difference between the plates is the same as the voltage of the battery. As long as we stay away from the edges of the plates, the electric field between the plates is uniform—both in magnitude and in direction. If the distance between the plates is d, the work performed in moving a charge q between the plates is

$$W = Fd = qEd$$

The electric potential difference between the plates is the work per unit charge. Therefore,

$$V = \frac{W}{q} = Ed$$

We can now solve for the magnitude of the electric field.

$$E = \frac{V}{d}$$

Example 20.4.2

Two large metal plates are separated by a distance d = 0.8 m and connected to a 6000-V battery as shown at the right. A gun fires an electron through a hole in the positive plate, and we find that the electron barely reaches point A exactly halfway between the plates before it stops and turns around. Find the electric field at point A and the initial velocity of the electron.

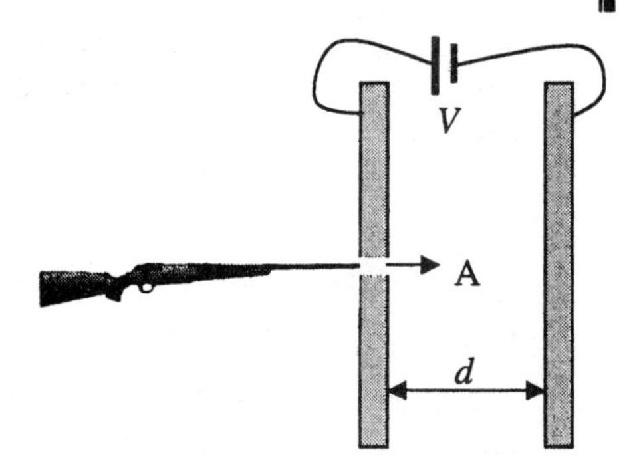

The electric field at point A is the same as the electric field at any other location between the two plates. It points to the right (from the positive plate towards the negative plate) and has the magnitude

$$E = \frac{V}{d} = \frac{6000 \text{ V}}{0.8 \text{ m}} = 7500 \text{ N/C}$$

We could use this electric field to find the force slowing the electron and then use the kinematic equations to solve for the initial velocity of the electron, but it is much easier to use conservation of energy. The electron loses all its kinetic energy as it travels from the positive plate to point A. Let's choose the electric potential energy of the electron to be zero at the positive plate.

$$\tfrac{1}{2}mv_o^2 + 0 = 0 + qE\left(\tfrac{1}{2}d\right)$$

$$v_o = \sqrt{\frac{qEd}{m}} = \sqrt{\frac{\left(-1.6\times10^{-19} \text{ C}\right)\left(-7500 \text{ N/C}\right)\left(0.8 \text{ m}\right)}{9.11\times10^{-31} \text{ kg}}} = 3.25\times10^{7} \text{ m/s}$$

Practice: The right plate is now moved closer so that the distance between the plates is only 0.4 m. If the gun fires the electron the same as before, will the electron reach the negative plate?

Answer: No. The electron is stopped by a potential difference of −3000 V. The potential difference between the plates will always be −6000 V as long as they are connected to the battery.

Problems

1. A penny initially contains 1.76×10^{23} protons and an equal number of electrons. The penny is then charged by removing 10^{15} electrons. What is the final net charge on the penny?
2. For the penny in the previous problem, what fraction of the original number of electrons would have to be removed to give it a net charge of +5 mC?
3. A nucleus contains 92 protons and 143 neutrons. What electric force does this nucleus exert on a single electron orbiting the nucleus at a distance of 0.1 nm?
4. If an atom has a diameter of 0.3 nm, what is the electric force between two electrons on opposite sides of the atom?
5. What is the electric force between two electrons separated by 1 m?
6. What is the electric force between two protons separated by 5×10^{-11} m? How does this compare to that between an electron and a proton separated by the same distance?
7. Equal charges of 2 C are placed at the corners of a square with sides 2 m long. What is the net force on each charge?
8. Equal charges of 2 C are placed at the corners of a right triangle whose short sides are each 1 m long. What is the net force on the charge at the right angle?
9. What is the acceleration of an electron at a distance of 3 cm from a charge of +2 mC?
10. What is the centripetal acceleration of an electron orbiting a proton at a distance of 0.05 nm?

*11. Show that Earth and the Moon would each need an electric charge of about 10^{14} C to have an electric force between them equal to their gravitational attraction.

*12. What equal charges would be needed for two 80-kg people to have the same electric repulsion as their gravitational attraction?

13. An electron experiences a force of 1.5×10^{-15} N directed north. What is the electric field experienced by the electron?

14. At a certain point in space, a charge $Q_1 = -6 \times 10^{-10}$ C experiences a net electric force of 1.2×10^{-4} N directed north. If Q_1 is removed and replaced by $Q_2 = 1.5 \times 10^{-9}$ C, what force will Q_2 experience?

15. At what distance from a charge $q = -3 \times 10^{-6}$ C would a charge $Q = +6 \times 10^{-6}$ C experience an electric field whose strength is 5×10^4 N/C?

16. Two pith balls are charged to $Q_A = -4.0 \times 10^{-9}$ C and $Q_B = +7 \times 10^{-9}$ C. With the pith balls held 0.23 m apart, what electric field is experienced by ball A (express the direction as either toward or away from ball B)?

17. Charges $Q_1 = 6 \times 10^{-8}$ C and $Q_2 = -6 \times 10^{-8}$ C are held in place at a separation of 0.16 m. What is the magnitude of the electric field midway between the two charges?

18. Charges $Q_1 = 5 \times 10^{-9}$ C and $Q_2 = 3 \times 10^{-9}$ C are separated by 0.1 m. What is the magnitude of the electric field midway between the two charges?

*19. What is the acceleration of an electron located 10^{-12} m from a fixed proton?

*20. A small ball has a charge of 1 mC and a mass of 100 g. What electric field would be needed to keep the ball suspended in the air?

21. It requires 30 J of work to move a charged object between two locations. If the electric potential energy of the object was originally 50 J, what is its potential energy at the new location?

22. It requires 75 J of work to move a charged object between two locations separated by 2 m that have an electric potential difference of 300 V. What is the charge on the object?

23. A 12-V battery does 48 J of work in pushing 4 C of charge through a circuit containing one light bulb. How much work is done by the same battery when it moves the same amount of charge through a circuit containing two bulbs in series?

24. Points A and C each have an electric potential of +6 V, and point B has an electric potential of +18 V. How much work is required to take 2 C of charge from A to B to C?

25. Points A and B have electric potentials of 5 V and 20 V, respectively. How much work would be required to take 3 C of positive charge from A to B? How much work would it take for 3 C of negative charge?

26. How much work does a 12-V battery do when one electron moves from the negative terminal to the positive terminal?

27. How far can a spark jump in dry air if the electric potential difference is 8 million V? (This is the reason high voltage sources are surrounded by a vacuum or an insulating fluid.)

28. What was your electric potential relative to a metal pipe if a spark jumped 8 cm (3 in.) from your finger to the pipe?

*29. Show that the units (V/m) used in describing the electric fields required for sparks and lightning are equivalent to those (N/C) used in the defining equation.

30. Show that the units (CV) used in conjunction with the work performed by a battery are the same as those (J) we used in Chapter 7.

31. Parallel plates are separated by 5 cm. The electric field between the plates is 7500 N/C. What is the potential difference between the plates?

32. The terminals of a 12-V battery are connected to parallel plates. What plate separation would produce an electric field between the plates of 1200 N/C?

21 — ELECTRIC CHARGES IN MOTION

21.1 Resistance

The resistance of a piece of wire depends on the type of wire, increases as the wire gets longer, and decreases as the wire is made bigger in diameter. These qualitative statements can be made quantitative by conducting measurements on different wires. The conclusions can be expressed by

$$R = \rho \frac{L}{A}$$

where R is the resistance, ρ is the **resistivity** of the material, L is the length, and A is the cross-sectional area of the wire. For a circular wire the cross-sectional area is given by $A = \pi r^2$.

The resistivity is a measure of the difficulty for an electric current to flow through a standard-sized piece of the material. The smaller the number, the larger the current for the same voltage. The resistivity is numerically equal to the resistance of a piece of the material that is 1 m long and 1 m^2 in cross-section. The unit of resistivity is an ohm-meter (Ω·m), as can be seen from the equation above. The values of the resistivities of some common materials are given in Table 21.1.1.

Table 21.1.1 Resistivities of Some Common Materials

Material	Resistivity (Ω·m)
Silver	1.59×10^{-8}
Copper	1.70×10^{-8}
Gold	2.44×10^{-8}
Aluminum	2.82×10^{-8}
Tungsten	5.6×10^{-8}
Nichrome	1.50×10^{-6}
Carbon	3.5×10^{-5}
Glass	10^{10} - 10^{14}
Hard Rubber	about 10^{13}
Amber	5×10^{14}

Example 21.1.1

What is the resistance of a piece of copper wire that has a diameter of 2 mm and a length of 5 m?

The cross-sectional area of the wire is given by

$$A = \pi r^2 = (3.14)\left(1\times10^{-3}\ \text{m}\right)^2 = 3.14\times10^{-6}\ \text{m}^2$$

The resistance is then

$$R = \rho\frac{L}{A} = \left(1.7\times10^{-8}\ \Omega\cdot\text{m}\right)\frac{5\ \text{m}}{3.14\times10^{-6}\ \text{m}^2} = 0.0271\ \Omega$$

Practice: What is the resistance of a tungsten wire used as the filament of a light bulb if it has a cross-sectional area of 10^{-9} m^2 and a length of 0.1 m?
Answer: 5.6 Ω

The resistances of the wires connecting light bulbs and other appliances are very much less than the resistances of the appliances. This means that one usually ignores the connecting wires when discussing the resistance of a typical circuit.

We can now calculate the total resistance of individual resistances connected in series or parallel. Series is the easiest, because this is just like putting two wires end to end. Because resistance depends on the length of wire, it is the total length that matters; resistances in series add.

$$R_t = R_1 + R_2$$ *resistances in series add*

where R_t is the total resistance and R_1 and R_2 are the individual resistances.

The rule for obtaining the total resistance of two resistors in parallel is a bit harder. We start by realizing that two wires in parallel means more area for the current, which in turn means an increase in the current. Let's look at the special case of two wires of equal length L. A simple assumption is that the combined resistance depends on the sum of the two areas.

$$A_t = A_1 + A_2$$

Using our resistivity equation, we can substitute for each area to obtain

$$\frac{\rho L}{R_t} = \frac{\rho L}{R_1} + \frac{\rho L}{R_2}$$

Canceling the common factor of ρL, we get

$$\frac{1}{R_t} = \frac{1}{R_1} + \frac{1}{R_2}$$ *resistances in parallel add as reciprocals*

Although this equation was obtained for a specific case of two wires of the same material and length, it actually holds for all resistances in parallel.

Example 21.1.2

What are the total resistances of a 4-Ω resistor and a 12-Ω resistor connected in series and parallel?

When they are connected in series, we have

$$R_t = R_1 + R_2 = 4\ \Omega + 12\ \Omega = 16\ \Omega$$

If we now connect them in parallel,

$$\frac{1}{R_t} = \frac{1}{R_1} + \frac{1}{R_2} = \frac{1}{4\,\Omega} + \frac{1}{12\,\Omega} = \frac{3}{12\,\Omega} + \frac{1}{12\,\Omega} = \frac{4}{12\,\Omega}$$

$$R_t = 3\,\Omega$$

Note that the total resistance of a parallel arrangement is less than either of the resistances alone. This is always true and makes sense because the second pathway will always make it easier for the electric charge to flow. No matter how large the second resistor, its addition into the circuit will always allow more charge to flow in a given amount of time.

Practice: What are the resistances of 2 Ω and 10 Ω connected in series and in parallel?
Answer: 12 Ω in series and 1.67 Ω in parallel

21.2 Ohm's Law

In the text we learned that resistance R is defined as the voltage V across an object divided by the current I through the object:

$$R = \frac{V}{I}$$

This equation is most useful when the resistance is constant, or nearly constant.

Example 21.2.1

A 12-Ω resistor and a 4-Ω resistor in series are connected to a 12-V battery. What is the current in the circuit and the voltage drop across each resistor?

Because resistances in series add, the battery sees a total resistance of 16 Ω. The current in the circuit is then calculated from Ohm's law.

$$I = \frac{V}{R} = \frac{12\text{ V}}{16\,\Omega} = \frac{3}{4}\text{ A}$$

Because this current must flow through each resistor, we can apply Ohm's law to each resistor to find the voltage drop across it.

$$V_1 = IR_1 = \left(\tfrac{3}{4}\text{ A}\right)(12\,\Omega) = 9\text{ V}$$

$$V_2 = IR_2 = \left(\tfrac{3}{4}\text{ A}\right)(4\,\Omega) = 3\text{ V}$$

As a check, notice that the total voltage drop is just 9 V + 3 V = 12 V.

Practice: If the two resistors are connected in parallel, what is the current through each?

Answer: 1 A through the 12-Ω resistor and 3 A through the 4-Ω resistor

Ohm's law provides us with an alternate method for obtaining the rules for combining resistances. To see this, let's start with two resistors R_1 and R_2 connected in series as shown in Figure 21.3.1. Conservation of energy requires that the voltage drop V_t across the pair be equal to the sum of the voltage drop V_1 across the first resistor and the voltage drop V_2 across the second resistor.

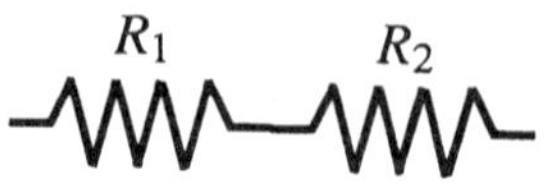

Fig. 21.3.1

$$V_t = V_1 + V_2$$

We can now use Ohm's law in the form $V = IR$ to substitute for the voltages.

$$I_t R_t = I_1 R_1 + I_2 R_2$$

Conservation of charge requires that the current through the combination be equal to that through each of the resistors. Therefore, $I_t = I_1 = I_2$ and we can cancel the currents to obtain

$$R_t = R_1 + R_2$$

resistors in series add

We can perform a similar analysis for the parallel resistors in Figure 21.3.2. In this case, we know that the total current through the pair must just be the sum of the currents through each one.

$$I_t = I_1 + I_2$$

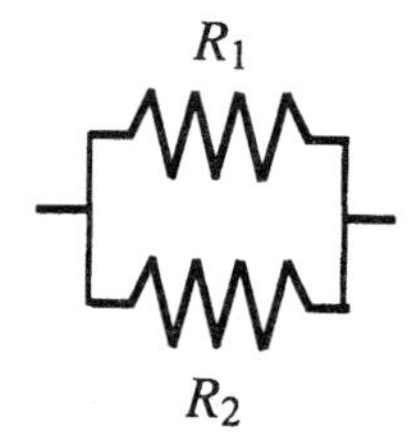

Fig. 21.3.2

Substituting for the currents using Ohm's law, we have

$$\frac{V_t}{R_t} = \frac{V_1}{R_1} + \frac{V_2}{R_2}$$

Because the voltage drop across the circuit is independent of the path, we know that $V_t = V_1 = V_2$. Therefore, we can cancel the voltages to obtain our result.

$$\frac{1}{R_t} = \frac{1}{R_1} = \frac{1}{R_2}$$

resistors in parallel add as reciprocals

21.3 Batteries and Bulbs

A light bulb is not an ohmic device; that is, its resistance is not constant. The hotter the filament, the greater the resistance. In spite of this complexity, circuits containing identical bulbs can provide great insight into the flow of electric charge through multiloop circuits. If one of two identical bulbs is glowing more brightly, the brighter bulb must have more charge flowing through its filament each second ("more flow, more glow"). Measurements show that the brighter bulb also has a greater voltage across its filament ("more volt, more jolt"). We can use these two principles to predict the relative brightnesses of identical bulbs in multiloop circuits.

Example 21.3.1

The circuit shown at the right contains 4 identical bulbs. Rank the bulbs according to their brightnesses.

This task could be stated another way: "Rank the bulbs according to the currents through them." If we think of the current through the battery as a river and follow the river as it flows from one end of the battery to the other, we see that bulb B only gets part of the total flow. What current doesn't flow through bulb B will flow through bulb A and then through bulb C. Because the paths rejoin, all of the current flows through bulb D, so bulb D is the brightest bulb. Which is second brightest? When the river divides between the path containing bulb B and the path containing bulbs A and C, does it divide 50-50, or some other way? Another guiding principle is that current "favors" the path of least resistance. It is important to use the word *favors* here instead of the word *chooses*. Some current will flow along each path, unless one of the paths has zero resistance. Because the path containing bulbs A and C has more resistance, this path will receive less than half of the current. Bulb B will therefore be brighter than bulb A. The current does not get "used up" in bulb A, but rather passes through to bulb C. Bulb C is therefore the same brightness as bulb A. The final ranking is: $D > B > A = C$.

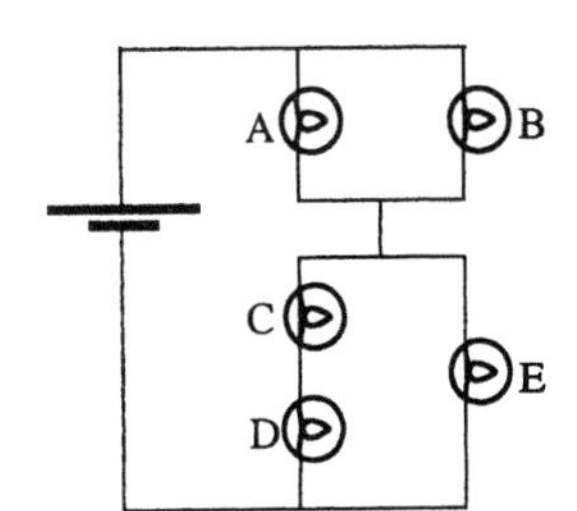

Practice: Rank the identical bulbs in the circuit shown at the right.
Answer: $E > A = B > C = D$

Whenever a new path is added to a circuit (added in parallel), no matter how resistive, the current through the battery will increase, indicating that the total resistance of the circuit has decreased. On the other hand, whenever a new resistance is added to an existing path (added in series), the total resistance of the circuit increases and the current through the battery decreases.

Example 21.3.2

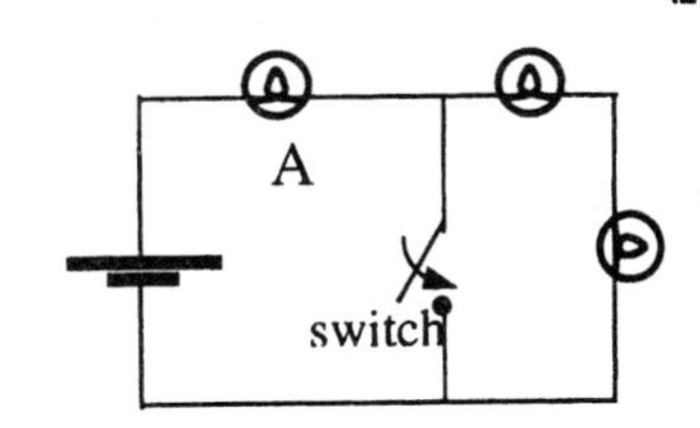

In the circuit shown at the right, the switch is initially open. When the switch is closed does the voltage across bulb A increase, decrease, or stay the same?

Closing the switch adds a new path—a new opportunity for flow in the circuit. The total resistance of the circuit decreases and the current through the battery increases. Because all the current passes through bulb A, bulb A must get brighter. A brighter bulb has more voltage across it, so the voltage across bulb A must increase.

Practice: What happens to the voltage across the other two bulbs when the switch is closed?
Answer: There is now a path back to the battery that has zero resistance. All of the current through bulb A will choose this path. The other two bulbs are *shorted out*. They go out and the voltage across them is zero.

Example 21.3.3

In the circuit shown at the right, switch S is initially open. When the switch is closed, does bulb B get brighter, dimmer, or stay the same?

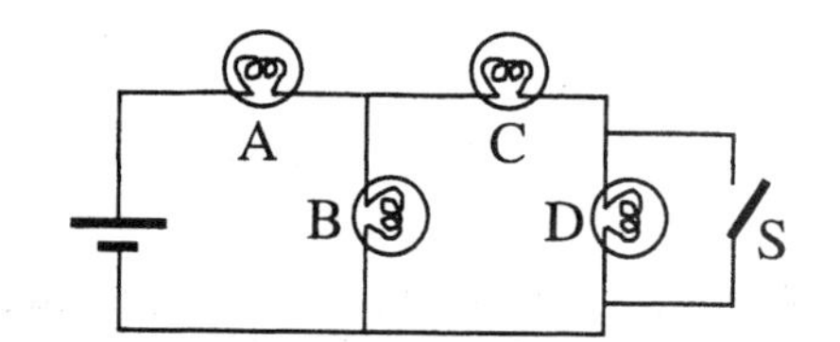

This example is more challenging. When the switch is closed, a path is added to the circuit, reducing the total resistance and increasing the current through the battery. Because all the current flows through bulb A, bulb A brightens. Bulb B shares the total current with the path containing bulb C. Closing the switch to form the new path changes the way the current is shared. Bulb B used to be on the path of least resistance, and therefore got more than half of the current. After the switch is closed, bulb D is shorted out and each path contains a single bulb. Therefore, bulb B gets half of the current. Bulb B gets a smaller share of the larger total current. Which effect dominates? The answer becomes apparent when we use the concept of voltage. The sum of the voltage drops around any path from one end of the battery to the other must be equal to the battery's voltage. Because bulb A gets brighter, the voltage across bulb A must increase and the voltage across bulb B must decrease. Therefore bulb B dims when the switch is closed.

Practice: What happens to the brightness of bulb C when the switch is closed?
Answer: Bulb C gets brighter.

21.4 Electric Power

The power P generated or dissipated in a device can be calculated from knowledge of the current through it and the voltage drop across it.

$$P = IV$$

For a device obeying Ohm's law, we can obtain other equivalent forms for the power. Substituting for the voltage, we get

$$P = IV = I\left(IR\right) = I^2R$$

Substituting for the current, we get

$$P = IV = \left(\frac{V}{R}\right)V = \frac{V^2}{R}$$

Example 21.4.1

What is the resistance of a 60-W bulb if it is rated for a voltage of 120 V?

Because we know the power and the voltage, we can use the last equation to find the resistance. Solving for the resistance R and plugging in the given numerical values, we find that

$$R = \frac{V^2}{P} = \frac{(120\text{ V})^2}{60\text{ W}} = 240\ \Omega$$

Practice: What is the resistance of a 100-W bulb?
Answer: 144 Ω

Example 21.4.2

The 60-W bulb and the 100-W bulb in Example 21.4.1 are now connected in series and plugged into the 120-V outlet. What is the power dissipated by each one?

When the bulbs are connected in series, their resistances add. Using the values obtained in Example 21.4.1, the total resistance is 384 Ω. This allows us to calculate the current through the circuit.

$$I = \frac{V}{R} = \frac{120\text{ V}}{384\ \Omega} = 0.313\text{ A}$$

Because this is also the current through each bulb, we can calculate the power for each bulb.

$$P_{60} = I^2 R_{60} = (0.313\text{ A})^2 (240\ \Omega) = 23.5\text{ W}$$

$$P_{100} = I^2 R_{100} = (0.313\text{ A})^2 (144\ \Omega) = 14.1\text{ W}$$

where the subscript refers to the wattage of the bulb. Notice that the power dissipated by each bulb is reduced from its labelled rating. This occurs because the voltage across each bulb is less than the 120 V normally used. What is surprising is that the power dissipated in the 100-W bulb is now less than that in the 60-W bulb. For the same currents, more power is dissipated in the bulb with the higher resistance according to $P = I^2R$.

Problems

1. How long must a copper wire be to have a resistance of 1 Ω if it has a square cross-section 1 mm on a side?

2. What is the resistance of a gold wire with a diameter of 0.2 mm and a length of 8 cm?

*3. A 2-m length of wire has a resistance of 16 Ω. The wire is cut in half and the two pieces are fastened side by side to make a 1-m length of thicker wire. What is the resulting resistance?

*4. A piece of copper wire has a resistance of 1 Ω. The wire is stretched uniformly to twice its original length, which of course reduces the cross-sectional area. What is the resistance of the stretched wire?

5. What is the total resistance of a 100-Ω resistor and a 1-Ω resistor connected in parallel? Does your answer agree with the statement that the total resistance is always smaller than the smaller resistance?

6. What is the total resistance of three 12-Ω resistors connected in series? In parallel?

7. What is the total resistance of the arrangement of resistors in Fig. 21.P.1?

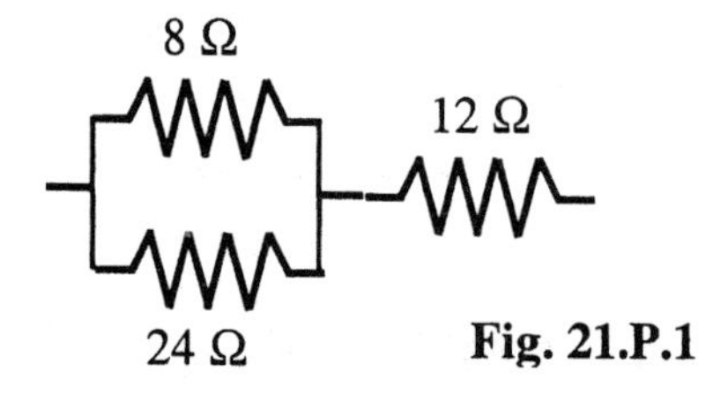

Fig. 21.P.1

8. What is the total resistance of the combination of resistors shown in Fig. 21.P.2?

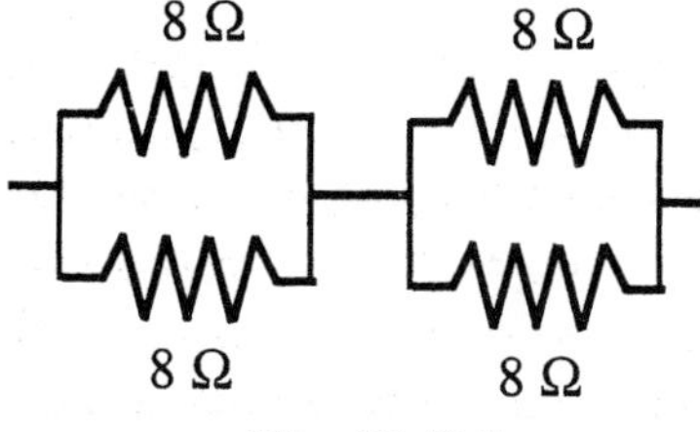

Fig. 21.P.2

*9. What different values of resistance can you get by combining three 6-Ω resistors?

*10. What values of resistance can you make by wiring four 8-Ω resistors together?

11. A resistor draws 6 A when connected across a 12-V battery. What additional single resistor could you add to the circuit to decrease the current through the battery to 1.5 A? Should you add it in series or in parallel?

12. A resistor draws 2 A when connected across a 12-V battery. What additional single resistor could you add to the circuit to increase the current through the battery to 8 A? Should you add it in series or in parallel?

13. Two resistors (4 Ω and 8 Ω) are connected in series to a 6-V battery. What is the voltage drop across each resistor?

14. Three resistors are connected in series to a 12-V battery. If the resistances are 4 Ω, 8 Ω, and 12 Ω, what is the voltage drop across each resistor?

15. Two resistors are connected in parallel to a 12-V battery. If the resistances are 8 Ω and 24 Ω, what is the current through each one?

16. If three 36-Ω resistors are connected in parallel to a 12-V battery, what is the total current supplied by the battery?

17. A battery is connected to a 2-Ω and a 4-Ω resistor connected in parallel. The current through the battery is 3 A. Find the equivalent resistance of the circuit, the battery voltage, and the current in the 2-Ω resistor?

*18. A 2-Ω resistor is connected in series with a network consisting of a 3-Ω and a 7-Ω resistor in parallel. If the current through the 2-Ω resistor is 5 A, find the current through the 3-Ω resistor.

19. If a string of Christmas tree lights has 24 bulbs, what voltage rating should each bulb have if they are wired in series?

20. What voltage rating should the bulbs in the previous problem have if they are wired in parallel?

21. The bulbs in the circuit in Fig 21.P.3 are identical. The switch S is closed. Rank the bulbs according to brightness.

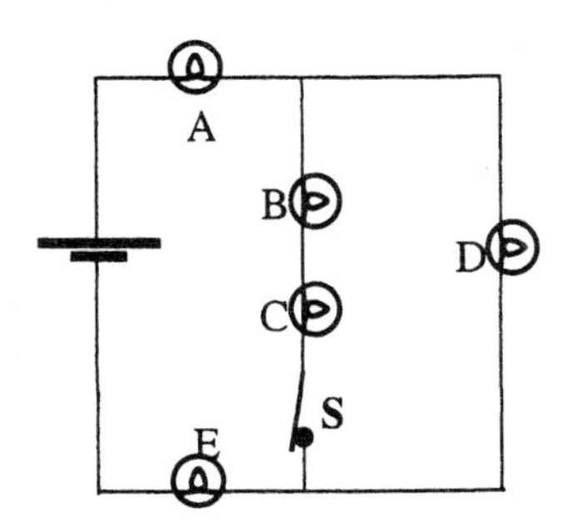

Fig. 21.P.3

22. The switch in the circuit shown in Fig. 21.P.3 is opened. Does the voltage across bulb D increase, decrease or stay the same?

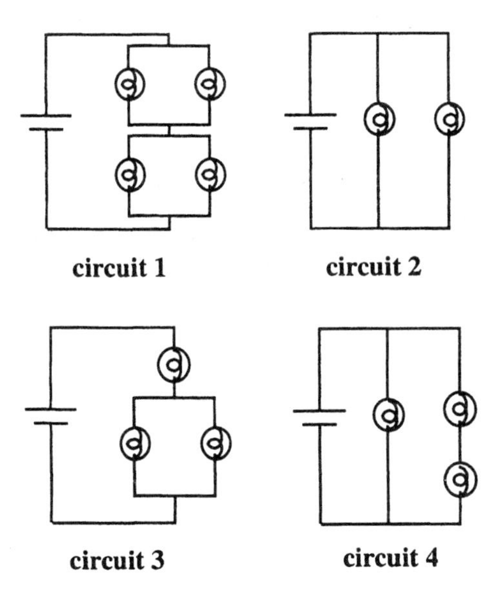

Fig 21.P.4

23. Which of the circuits in Fig. 21.P.4 has the least equivalent resistance?

24. Which of the circuits in Fig. 21. P.4 has the most equivalent resistance?

25. Identical bulbs are wired as shown in the circuit in Fig. 21.P.5. Rank the bulbs from brightest to dimmest. Explain your reasoning.

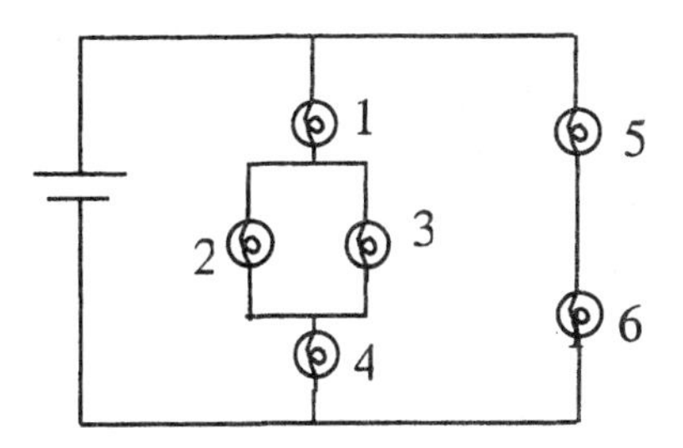

Fig. 21.P.5

26. If bulb 2 in the circuit in Fig. 21.P.5 is removed from its socket (but the empty socket remains), will bulb 1 get brighter, stay the same, or get dimmer? What about bulb 3?

*27. The switch shown in Fig. 21.P.6 can be used to connect the wire on the left to either of those on the right. Design a circuit using two of these switches to independently turn a light bulb on and off like the switches at each end of a hallway. Be sure that there are no short circuits when the bulb is off.

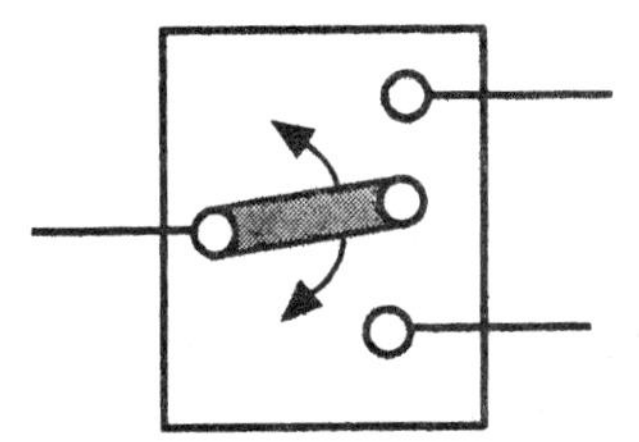

Fig. 21.P.6

*28. The switch shown in Fig. 21.P.7 can be used to reverse the connections between the two wires on the left and the two on the right. Add one of these switches to the circuit in the previous question so that three switches control the same bulb.

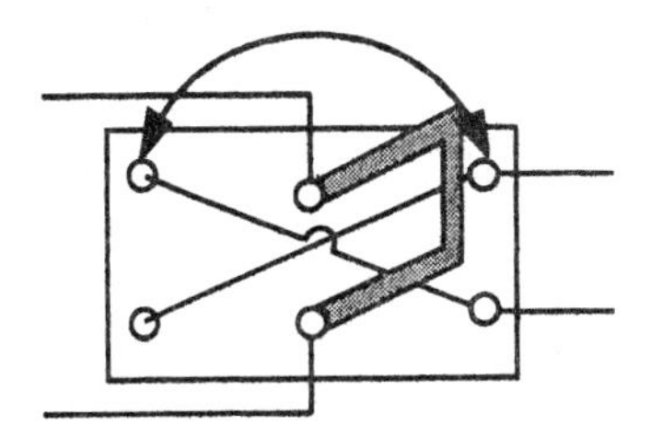

Fig. 21.P.7

29. What power is need to operate a clothes dryer with a resistance of 12 Ω?

30. A dishwasher has a resistance of 10 Ω. What power does it require?

31. A toaster is rated at 800 W. If a slice of bread is toasted in 120 s, how much energy is used?

32. How much energy is required to run a 1500-W space heater for a 24-hour day?

33. At a cost of 8¢/kWh, what does it cost to leave a 150-W security light on during a 10-hour night?

34. What does it cost to run a 1200-W space heater for 12 hours if electrical energy costs 8¢/kWh?

35. How many 100-W bulbs can be put in parallel on one 120-V circuit before they blow a 20-A fuse?

36. What size fuse is needed if you want to run two 1500-W space heaters on the same circuit?

37. Two 6-Ω resistors are connected in parallel to a 12-V battery. What is the electrical power used? How does this compare to a single 6-Ω resistor connected to the 12-V battery?

38. Two 6-Ω resistors are connected in series to a 12-V battery. What is the electrical power used? How does this compare to a single 6-Ω resistor connected to the 12-V battery?

*39. If a 10-W and a 20-W bulb are connected in parallel, the 20-W bulb is brighter. Which one is brighter if they are connected in series? Explain.

22 — ELECTROMAGNETISM

22.1 Charged Particles in Magnetic Fields

The force exerted on a moving charged particle by a magnetic field is given by

$$F = qv_{\perp}B$$

where q is the particle's charge, B is the strength of the magnetic field, and $v_{\perp}$ is the component of the velocity that is perpendicular to the direction of the magnetic field. The direction of the force is always perpendicular to the direction of the magnetic field and to the direction of the velocity as shown in Figure 22.1.1.

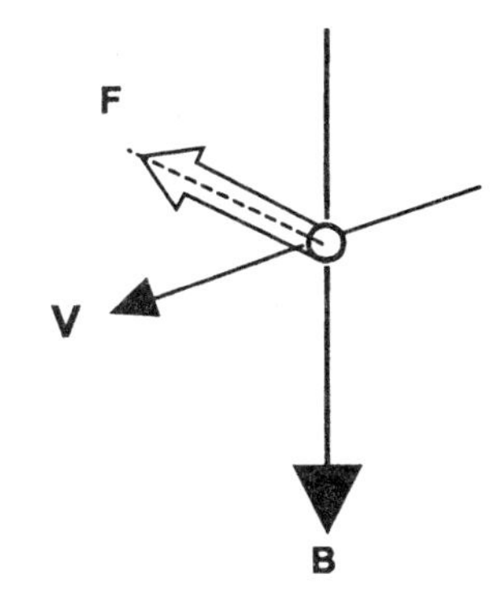

Fig. 22.1.1

For the special case when the velocity v is perpendicular to the magnetic field, the force has a constant value and always points at right angles to the instantaneous velocity. This is a centripetal force that will cause the particle to travel along a circular path. The centripetal acceleration a of the particle can be written

$$a = \frac{v^2}{r}$$

where r is the radius of the circular path. Because this centripetal acceleration is provided by the magnetic force, we can also write the acceleration as

$$a = \frac{F}{m} = \frac{qvB}{m}$$

where m is the particle's mass. Equating these two expressions for the acceleration, we have

$$\frac{v^2}{r} = \frac{qvB}{m}$$

Canceling the common factor v and solving for r, we arrive at our expression for the radius of the circle.

$$r = \frac{mv}{qB}$$

The units of the magnetic field are tesla (T). From our force equation at the top of the page, we can see that a tesla must have the following equivalent units.

$$\mathrm{T} = \frac{\mathrm{N}}{\mathrm{C \cdot m/s}} = \frac{\mathrm{kg \cdot m}}{\mathrm{s^2}}\frac{\mathrm{s}}{\mathrm{C \cdot m}} = \frac{\mathrm{kg}}{\mathrm{C \cdot s}} = \frac{\mathrm{kg}}{\mathrm{A \cdot s^2}}$$

Example 22.1.1

What is the acceleration of a proton with a velocity of 3×10^6 m/s perpendicular to a magnetic field with a strength of 0.4 T?

$$a = \frac{qvB}{m} = \frac{(1.6\times10^{-19}\ \text{C})(3\times10^{6}\ \text{m/s})(0.4\ \text{T})}{1.67\times10^{-27}\ \text{kg}} = 1.15\times10^{14}\ \text{m/s}^2$$

Practice: What is the acceleration if we replace the proton by an electron?
Answer: 2.11×10^{17} m/s^2

Example 22.1.2

What is the radius of the circular path for the proton in the previous example?

$$r = \frac{mv}{qB} = \frac{(1.67\times10^{-27}\ \text{kg})(3\times10^{6}\ \text{m/s})}{(1.6\times10^{-19}\ \text{C})(0.4\ \text{T})} = 7.38\times10^{-2}\ \text{m} = 7.83\ \text{cm}$$

Practice: What is the radius for the electron?
Answer: 4.27×10^{-5} m

Practice: Show that you can get the same radius using the results of Example 22.1.1 and the equation for the centripetal acceleration.

22.2 Faraday's Law

Whenever the number of magnetic field lines through a conducting loop is changed, a voltage is created, which can drive an induced current around the loop. In the special case where the loop is perpendicular to a uniform magnetic field, Faraday's law is written

$$V = \frac{\Delta(BA)}{\Delta t}$$

where A is the area of the loop and B is the strength of the magnetic field. The direction of the induced current is determined according to Lenz's law.

Example 22.2.1

A loop with an area of 0.2 m^2 and a resistance of 8 Ω is perpendicular to a constant magnetic field with a strength of 0.05 T. What is the induced current in the loop?

Because the magnetic field is not changing, there will be no induced voltage and therefore no induced current.

What is the induced current if the magnetic field decreases steadily to zero in a time of 10 s?

$$I = \frac{V}{R} = \frac{1}{R}\frac{\Delta(BA)}{\Delta t} = \frac{A}{R}\frac{\Delta B}{\Delta t} = \frac{(0.2\ \text{m}^2)(0.05\ \text{T})}{(8\ \Omega)(10\ \text{s})} = 0.125\ \text{mA}$$

Example 22.2.2

Two parallel copper rails are connected by a 20-Ω resistor at one end and are spaced 0.5 m apart. The rails support a copper bar that is free to slide along the rails as shown. A uniform magnetic field $B = 0.4$ T is directed perpendicular to the plane formed by the rails and into the page. If the copper bar is forced to slide along the rails to the right at a speed of 2 m/s, what induced current will flow through the resistor?

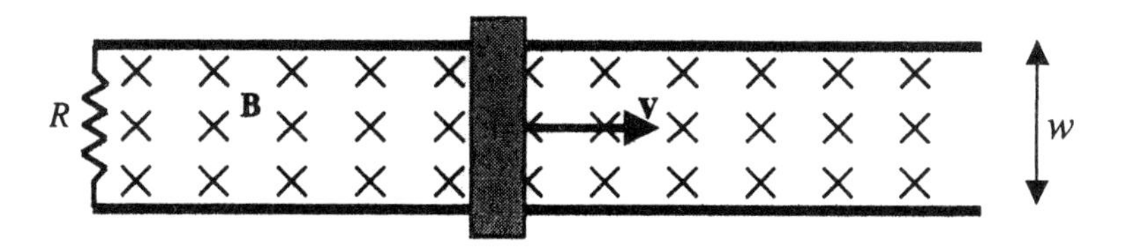

In this case the loop consists of the rails, the bar, and the connecting resistor. The magnetic field is not changing with time, but the number of field lines passing through the loop is increasing as the bar moves to the right. The magnitude of the induced current in the resistor (if we assume that the resistance of the copper rails and bar is much less than that of the resistor) is

$$I = \frac{V}{R} = \frac{1}{R}\frac{\Delta(BA)}{\Delta t} = \frac{B}{R}\frac{\Delta A}{\Delta t} = \frac{Bw}{R}\frac{\Delta x}{\Delta t} = \frac{Bwv}{R}$$

$$= \frac{(0.4\ \text{T})(0.5\ \text{m})(2\ \text{m/s})}{20\ \Omega} = 0.02\ \text{A}$$

22.3 Transformers

The relative size of the voltage produced in the secondary coil of a transformer is given by the ratio of the number of loops in the secondary to the number of loops in the primary. Using the subscripts p and s for primary and secondary, respectively, this gives the following relationship.

$$\frac{V_s}{V_p} = \frac{N_s}{N_p}$$

Example 22.3.1

A transformer is used to reduce the voltage from 120 V to 9 V. If the primary coil has 400 turns, how many coils must the secondary have?

Solving our transformer equation for the number of coils N_s in the secondary, we have

$$N_s = N_p \frac{V_s}{V_p} = (400 \text{ turns}) \frac{9 \text{ V}}{120 \text{ V}} = 30 \text{ turns}$$

Practice: If the transformer had 40 turns in the secondary coil, what would the output voltage be?
Answer: 12 V

22.4 Making Waves

Example 22.4.1

The range of frequencies for visible light is 4.0–7.5 × 10^{14} Hz. What is the wavelength corresponding to the reddest red light?

Remembering that red light has the lowest frequency, we can calculate the wavelength.

$$\lambda = \frac{c}{f} = \frac{3 \times 10^8 \text{ m/s}}{4.0 \times 10^{14}} = 7.5 \times 10^{-7} \text{ m} = 750 \text{ nm}$$

Practice: What is the wavelength of the most violet violet light?
Answer: 400 nm

Problems

1. An electron is traveling at 8×10^4 m/s perpendicular to a magnetic field of 4 T. What is the magnetic force on the electron?
2. What is the maximum force a magnetic field of 5 T can exert on a proton traveling at 4×10^6 m/s?
3. You find that the maximum force on a particle is 4×10^{-3} N when it passes through a magnetic field of strength 0.4 T at a speed of 120 m/s. What is the charge on the particle?
4. What minimum strength of magnetic field is required to balance the gravitational force on an electron moving at speed 3×10^6 m/s?
5. An electron traveling at 3×10^7 m/s perpendicular to a magnetic field of 0.2 T executes a circular path. What is the radius of the circle?

6. What is the radius of the circle followed by a proton with a velocity of 3×10^7 m/s perpendicular to a magnetic field of 0.3 T?

7. What magnetic field is required to constrain electrons moving at 6×10^6 m/s to a circle of radius 0.5 m?

8. What magnetic field is required to constrain protons moving at 6×10^6 m/s to a circle of radius 0.5 m?

9. A uniform magnetic field of strength 0.05 T is used in a cyclotron to constrain protons to circular orbits of radius 1 m. What is the speed of the protons? How long does it take a proton to complete one revolution?

10. In the cyclotron in the previous problem, a slower proton is moving in a circle of radius 0.5 m. How long does it take this proton to complete one revolution? Compare this to the answer for the previous problem.

11. A flexible loop of conducting material has a radius of 20 cm and is perpendicular to a uniform magnetic field of strength 0.3 T. You grab the loop at points A and B and pull them apart so that the area of the loop is reduced to zero in a time of 0.1 s. If the resistance of the loop is 3 Ω, find the average current in the loop during this time.

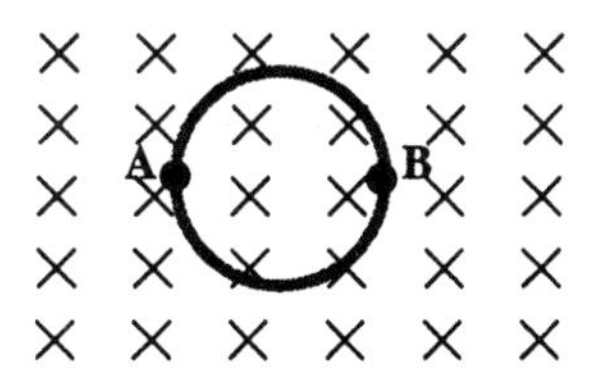

12. A "smart light" is triggered by changing the strength of the magnetic field through a loop of wire embedded in the pavement. The loop is typically a rectangle of area 2 m^2 and has a resistance of 0.1 Ω. Before your car arrives at the loop, the magnetic field through the loop is due to Earth's magnetic field, and in Montana the component perpendicular to the loop is 4.7×10^{-5} T. If the average strength of this field through the loop is tripled by the presence of your iron engine block, find the average current through the loop during the 2 s it takes to drive onto the loop.

13. How fast would you have to slide the copper bar in Example 22.2.2 to induce a current of 0.5 A in the resistor?

*14. The power dissipated in the resistor of the previous problem must be supplied by an external agent pushing on the copper bar to make it slide. Find the constant force exerted by the external agent on the bar. If the bar is moving at a constant velocity, how does the magnitude of this external force compare to the magnitude of the magnetic force acting on the bar as it slides through the magnetic field?

*15. A piece of copper wire is formed into a single circular loop with an area of 0.2 m^2. A uniform magnetic field is oriented perpendicular to the plane of the loop. The strength of this field increases steadily from 0 to 0.7 T in a time of 0.4 s. Find the energy dissipated in the wire if it has resistance 0.03 Ω.

*16. How does your answer to the previous problem change (if at all) if the same change in magnetic field takes place in half the time?

17. A transformer designed to reduce the voltage from 24 kV to 120 V has 750 turns in the secondary coil. How many turns does it have in the primary?

18. A transformer designed to reduce the voltage from 120 V to 12 V has 1500 turns in the primary. How many turns does it have in the secondary?

19. What output voltage would you expect if the primary has 500 turns, the secondary has 50 turns and the input voltage is 120 V ac?

20. What output voltage would you expect in the previous problem if the input voltage were 120 V dc?

21. What is the frequency of an X ray with a wavelength of 0.08 nm?

22. Typical microwaves have a wavelength of 12.2 cm. What is the frequency of these microwaves?
23. What is the frequency of the "X-band" radar used by police if it has a wavelength of 2.85 cm?
24. The eye is most sensitive to yellow-green light with wavelengths around 550 nm. What is the frequency of this light?
25. If an FM radio station broadcasts at 100.7 MHz, at what wavelength is it broadcasting?
26. The emergency frequency for airplanes is 121.1 MHz. What is the wavelength of these radio waves?
27. Channel 2 on your TV set broadcasts in a band from 54 to 60 MHz. What is the corresponding range of wavelengths?

23 — THE EARLY ATOM

23.1 The Discovery of the Electron

With the determination of the sizes of the charge and mass of the electron, we can now perform some calculations to try to develop a feeling for the sizes of these quantities. We begin by calculating how many electrons it would take to have a total mass of 1 kg. To determine how many it takes, we divide the total mass by the mass of each electron. Thus

$$\# = \frac{1\text{ kg}}{9.11\times10^{-31}\text{ kg/electron}} = 1.1\times10^{30}\text{ electrons}$$

This is a very large number—one million million million million million electrons.

Similarly we can find the number of electrons that we would have to gather together to have 1 C of electric charge.

$$\# = \frac{1\text{ C}}{1.6\times10^{-19}\text{ C/electron}} = 6.25\times10^{18}\text{ electrons}$$

And, finally, we can calculate the total mass of the electrons carrying this 1 C of charge.

$$m = \left(9.11\times10^{-31}\frac{\text{kg}}{\text{electron}}\right)\left(6.25\times10^{18}\text{ electrons}\right) = 5.69\times10^{-12}\text{ kg}$$

23.2 Radiating Objects

Example 23.2.1

What frequency is needed to produce quanta with energies of 1 eV?

The electron volt (eV) is a unit of energy customarily used at the atomic and subatomic levels. It is the kinetic energy acquired by an electron or proton falling through a potential difference of 1 V. From its definition we know that

$$1\text{ eV} = \left(1.6\times10^{-19}\text{ C}\right)\left(1\text{ V}\right) = 1.6\times10^{-19}\text{ J}$$

Therefore, using this value as a conversion factor, we find the frequency.

$$f = \frac{E}{h} = \frac{1\text{ eV}}{6.63\times10^{-34}\text{ J}\cdot\text{s}}\left[\frac{1.6\times10^{-19}\text{ J}}{1\text{ eV}}\right] = 2.41\times10^{14}\text{ Hz}$$

According to Figure 22-26 in the text, this frequency is in the infrared range.

Practice: What frequency is needed to produce quanta with twice the energy?
Answer: Double the frequency

Example 23.2.2

What are the energies (in eV) of the bluest blue and reddest red photons of the visual spectrum?

$$E_{red} = hf_{red} = \left(6.63\times10^{-34}\ \text{J}\cdot\text{s}\right)\left(4\times10^{14}\ \text{Hz}\right) = 2.65\times10^{-19}\ \text{J}\left[\frac{1\ \text{eV}}{1.6\times10^{-19}\ \text{J}}\right] = 1.66\ \text{eV}$$

$$E_{blue} = hf_{blue} = \left(6.63\times10^{-34}\ \text{J}\cdot\text{s}\right)\left(7.5\times10^{14}\ \text{Hz}\right) = 4.97\times10^{-19}\ \text{J}\left[\frac{1\ \text{eV}}{1.6\times10^{-19}\ \text{J}}\right] = 3.11\ \text{eV}$$

23.3 The Photoelectric Effect

The description of the photoelectric effect described in the text can be translated into an equation using the conservation of energy.

$$KE_{max} = hf - \phi$$

where KE_{max} is the maximum kinetic energy of the emitted electrons, hf is the energy of the incident photons, and ϕ is the amount of energy required for electrons to leave the surface. The reason that the equation involves the maximum kinetic energy is that it takes a range of energies for the electrons to get to the surface depending on their locations in the metal. There are many electrons with less than the maximum kinetic energy. The values of the *work function* ϕ for typical metallic surfaces range from 2 eV to 7 eV with most being between 4 eV and 5 eV. Because the kinetic energies of electrons are often expressed in eV, it is useful to write the value of Planck's constant in these units.

$$h = 6.63\times10^{-34}\ \text{J}\cdot\text{s}\left[\frac{1\,\text{eV}}{1.6\times10^{-19}\ \text{J}}\right] = 4.14\times10^{-15}\ \text{eV}\cdot\text{s}$$

Example 23.3.1

What is the maximum kinetic energy of the photoelectrons emitted when ultraviolet light with a frequency $f = 2\times10^{15}$ Hz is shined on an aluminum surface with a work function $\phi = 4.08$ eV?

$$KE_{max} = hf - \phi = \left(4.14\times10^{-15}\ \text{eV}\cdot\text{s}\right)\left(2\times10^{15}\ \text{Hz}\right) - 4.08\ \text{eV} = 4.2\ \text{eV}$$

Practice: What is the maximum kinetic energy for twice the frequency?
Answer: 12.5 eV

Example 23.3.2

What is the minimum frequency that will eject photoelectrons from aluminum?

We set the kinetic energy equal to zero to find out the lowest energy photon that can eject an electron. Notice that this photon must only supply enough energy for the electron to escape the surface. With $KE_{max} = 0$, our equation shows that the photon energy is equal to the work function. Solving for the frequency, we get

$$hf = \phi$$

$$f = \frac{\phi}{h} = \frac{4.08 \text{ eV}}{4.14 \times 10^{-15} \text{ eV} \cdot \text{s}} = 9.86 \times 10^{14} \text{ Hz}$$

Example 23.3.3

Red light of wavelength 670 nm produces photoelectrons from a certain metal that leave with kinetic energies between 0 and 0.5 eV. What is the work function of the metal?

$$\varphi = hf - KE_{max} = \frac{hc}{\lambda} - KE_{max}$$

$$= \frac{(4.14 \times 10^{-15} \text{ eV} \cdot \text{s})(3 \times 10^{8} \text{ m/s})}{670 \times 10^{-9} \text{ m}} - 0.5 \text{ eV} = 1.35 \text{ eV}$$

Practice: What is the maximum wavelength that will eject photoelectrons from this metal?
Answer: 917 nm

23.4 Bohr's Model

Bohr's restriction on the allowed values of the angular momentum in the hydrogen atom allows the calculation of the values associated with each orbit. We begin by obtaining the values for the ground state; that is, for $n = 1$. The smallest possible angular momentum L_1 is given by

$$L_1 = \frac{h}{2\pi} = \frac{6.63 \times 10^{-34} \text{ J} \cdot \text{s}}{6.28} = 1.06 \times 10^{-34} \text{ J} \cdot \text{s}$$

The other values are given by $L_n = nL_1$.

The smallest radius r_1 is given by

$$r_1 = \frac{h^2}{4\pi^2 kme^2} = 5.29 \times 10^{-11} \text{ m} = 0.0529 \text{ nm}$$

where m is the mass of the electron, e is its charge, and k is Coulomb's constant. The other values of the radius are given by $r_n = n^2 r_1$.

The lowest energy is given by

$$E_1 = \frac{-ke^2}{2r_1} = -2.17\times10^{-18}\text{ J}\left[\frac{1\text{ eV}}{1.6\times10^{-19}\text{ J}}\right] = -13.6\text{ eV}$$

The energy has a negative value because of the choice made for the zero value of the potential energy. It is chosen to be zero when the electron is located an infinite distance from the nucleus of the atom. The negative value then tells us that the electron is bound to the nucleus and requires an energy equal to E_1 to escape. The other values of the energy are given by $E_n = E_1/n^2$. Notice that even though the energies of the larger orbits are obtained by dividing by n^2, the energies are higher because of the minus sign.

Example 23.4.1

What are the radius and the energy of the first excited state in hydrogen?

$$r_2 = n^2 r_1 = 2^2(0.0529\text{ nm}) = 0.212\text{ nm}$$

$$E_2 = \frac{E_1}{n^2} = \frac{-13.6\text{ eV}}{2^2} = -3.4\text{ eV}$$

23.5 Atomic Spectra Explained

The energies of the photons emitted when the electrons in hydrogen atoms drop from the mth level to the nth level can be calculated from our expression for the energies of the various levels.

$$E_m - E_n = \frac{E_1}{m^2} - \frac{E_1}{n^2} = E_1\left(\frac{1}{m^2} - \frac{1}{n^2}\right)$$

Notice that because $m > n$, the term in brackets will be negative and the energy will be positive as expected. It is these photons that comprise the spectral lines given off by hydrogen when the atoms are excited.

Example 23.5.1

What are the energy and the frequency of the photons emitted when electrons in hydrogen drop from the $n = 2$ level to the $n = 1$ level?

$$E_2 - E_1 = E_1\left(\frac{1}{2^2} - \frac{1}{1^2}\right) = \frac{-3E_1}{4} = 10.2\text{ eV}$$

We can now calculate the frequency of photons with this energy.

$$f = \frac{E}{h} = \frac{10.2\ \text{eV}}{6.63\times10^{-34}\ \text{J}\cdot\text{s}}\left[\frac{1.60\times10^{-19}\ \text{J}}{1\ \text{eV}}\right] = 2.46\times10^{15}\ \text{Hz}$$

These photons are in the ultraviolet range and are not visible to the naked eye.

Practice: What is the energy of the photons emitted in the n = 3 to n = 2 transition?
Answer: 1.89 eV

Problems

1. What energy quanta correspond to the broadcast frequency (96.7 MHz) of KSKY?
2. What energy quanta correspond to the frequency (5.5×10^{14} Hz) where the human eye is most sensitive?
3. What frequency would produce quanta with an energy of 1 J? Does this seem like a realistic frequency?
4. What frequency would produce quanta with an energy of 1 MeV?
5. What is the energy quanta corresponding to a wavelength of 1 m?
6. When an electron meets a positron (like an electron but with positive charge), the two particles annihilate with the release of two "511-keV" photons. What is the wavelength of these photons?
7. Light with a frequency of 2×10^{15} Hz shines on a copper surface (ϕ = 4.70 eV). What is the maximum kinetic energy of the photoelectrons?
8. What is the maximum kinetic energy of the photoelectrons emitted when light with a frequency of 5×10^{15} Hz shines on a zinc surface that has a work function of 4.31 eV?
9. What is the minimum frequency that will produce photoelectrons from a silver surface with a work function of 4.73 eV?
10. A polished surface of platinum (ϕ = 6.35 eV) is illuminated by light with a frequency of 1.7×10^{15} Hz. Will photoelectrons be emitted?
11. You shine ultraviolet light of wavelength 300 nm onto a metal surface and find that the resulting photoelectrons have a maximum kinetic energy 0.7 eV. What is the work function for this metal?
12. Light of frequency 1.6×10^{15} Hz is shone onto a metal surface. The resulting photoelectrons have a maximum kinetic energy of 2.5 eV. What is the work function for this metal?
13. What is the angular momentum of the electron in the second Bohr orbit?
14. Show that the units of Planck's constant are those of angular momentum.
15. What is the radius of the second excited state in hydrogen?
16. What is the radius of the n = 4 state in hydrogen?
17. What is the energy of the second excited state in hydrogen?
18. What is the energy of the n = 4 state in hydrogen?
19. How much energy is required to remove the electron from the hydrogen atom when it is in the first excited state?

20. What is the energy of the photon that is released when the electron drops from the $n = 4$ to the $n = 1$ level?

21. Is the photon given off in the previous problem in the visible, ultraviolet, or X-ray range of the electromagnetic spectrum?

22. What color is emitted when the electrons in hydrogen atoms drop from the $n = 4$ to the $n = 2$ level?

23. What is the frequency emitted when an electron is captured into the ground state of hydrogen?

24. What frequency is emitted when electrons in hydrogen drop from the $n = 3$ level to the $n = 1$ level?

25. What is the wavelength of light that results when electrons in a hydrogen atom drop from the $n = 5$ to the $n = 2$ level?

*26. A photon of energy 2.55 eV is observed as a hydrogen atom drops to a lower energy state. If the electron falls to the $n = 2$ level, which level did it come from?

24 — THE MODERN ATOM

24.1 De Broglie's Waves

Louis de Broglie showed that particles have wave properties as well as particle properties. The wavelength associated with these particles is given by

$$\lambda = \frac{h}{mv}$$

where h is Planck's constant, m is the mass of the particle, and v is the speed of the particle.

Example 24.1.1

What is the wavelength of an electron with a kinetic energy of 1 eV?

Because the de Broglie relationship depends on momentum, we need to find the momentum of the electron first. Starting with the definition of kinetic energy, and treating the electron nonrelativistically, we have

$$KE = \tfrac{1}{2}mv^2$$

If we multiply both sides of the equation by $2m$, we get

$$m^2v^2 = 2mKE = 2\left(9.11\times10^{-31}\text{ kg}\right)\left(1\text{ eV}\right)\left[\frac{1.6\times10^{-19}\text{ J}}{1\text{ eV}}\right] = 2.92\times10^{-49}\,\frac{\text{kg}^2\cdot\text{m}^2}{\text{s}^2}$$

Therefore

$$mv = 5.4\times10^{-25}\text{ kg}\cdot\text{m/s}$$

Putting this value in the de Broglie relationship, we have

$$\lambda = \frac{h}{mv} = \frac{6.63\times10^{-34}\text{ J}\cdot\text{s}}{5.4\times10^{-25}\text{ kg}\cdot\text{m/s}} = 1.23\times10^{-9}\text{ m} = 1.23\text{ nm}$$

This wavelength is about 10 times the spacing between atoms in a crystal.

Practice: What happens to the wavelength if the electron has 10 times the kinetic energy?
Answer: The wavelength decreases by the square root of 10 or a factor of 3.16. Therefore, λ = 0.389 nm.

Example 24.1.2

The resolving power of a microscope is approximately equal to that of the wavelength of the light used. In an electron microscope, electrons are used as the "light" source. What kinetic energy electrons are needed if the resolving power is required to be 10^{-11} m, which would enable one to "see" an atom?

We can solve de Broglie's relationship to find the required electron speed

$$v = \frac{h}{m\lambda} = \frac{6.63\times10^{-34}\ \text{J}\cdot\text{s}}{\left(9.11\times10^{-31}\ \text{kg}\right)\left(10^{-11}\ \text{m}\right)} = 7.28\times10^{7}\ \text{m/s}$$

As this speed is less than 25% of the speed of light, we use the classical formula to calculate the approximate kinetic energy.

$$E = \tfrac{1}{2}mv^2 = \tfrac{1}{2}\left(9.11\times10^{-31}\ \text{kg}\right)\left(7.28\times10^{7}\ \text{m/s}\right)^2\left[\frac{1\ \text{eV}}{1.6\times10^{-19}\ \text{J}}\right] = 15.1\ \text{keV}$$

24.2 A Particle in a Box

If a particle is confined to a one-dimensional box, the particle's wavelength must satisfy the same conditions as a standing wave on a guitar string; that is, the fundamental wavelength must be twice the length of the box. Therefore, the lowest value for the momentum of the particle is

$$p_1 = \frac{h}{\lambda_1} = \frac{h}{2L}$$

We know that a guitar string has all harmonics. By analogy the possible momenta of the particle are

$$p_n = np_1 \qquad n = 1, 2, 3, \ldots$$

If the box is long enough that we can treat the particle non-relativistically, we can also compute the possible kinetic energies of the particle.

$$KE_n = \tfrac{1}{2}mv_n^2 = \frac{p_n^2}{2m} = n^2\frac{h^2}{8mL^2} \qquad n = 1, 2, 3, \ldots$$

Note that the lowest energy allowed for a particle confined to a box is not zero. If the particle's energy could be zero, both its momentum and the uncertainty in its momentum would be zero. Therefore, the uncertainty in its position must be infinite. But we know that the particle is confined to the box in violation of the uncertainty principle.

We also note that it would be impossible to lower the temperature of a collection of such particles to zero in violation of the third law of thermodynamics.

Example 24.2.1

What is the lowest kinetic energy for an electron confined to a one-dimensional box with a length of 0.1 nm (about the diameter of the hydrogen atom)?

$$KE_1 = \frac{h^2}{8mL^2} = \frac{\left(6.63\times10^{-34}\ \text{J}\cdot\text{s}\right)^2}{8\left(9.11\times10^{-31}\ \text{kg}\right)\left(10^{-10}\ \text{m}\right)^2}\left[\frac{1\ \text{eV}}{1.6\times10^{-19}\ \text{J}}\right] = 37.7\ \text{eV}$$

Practice: What is the kinetic energy of the electron in the first excited state?
Answer: 151 eV

24.3 The Uncertainty Principle

The Heisenberg uncertainty principle puts limits on the simultaneous measurements of a particle's position and momentum. This relationship can be used to make estimates of a number of effects that will give us some insight into subatomic phenomena.

Example 24.3.1

A beam of electrons with a speed $v = 10^4$ m/s passes through a vertical slit with a width w = 20 μm. What is the width of the diffraction pattern produced on a screen located L = 1 m away?

Let's first estimate the answer using the uncertainty principle. We can do this because confining the electrons to pass through the slit gives an uncertainty in their positions in the horizontal direction and, thus, creates an uncertainty in their horizontal component of momentum. This means that even though the electrons were aimed face on to the slit, they can emerge from the other side with a sideways component of the momentum. This sideways component means that the electron can hit the screen to either side of the center line. Let's use the symbol y for the sideways displacement of the electrons.

$$\Delta p\,\Delta y > h$$

Let's set the uncertainty in the horizontal position Δy equal to $w/2$. To do calculations, we set the two sides of the relationship "approximately equal" to each other. Solving for Δp, we can find the uncertainty in the component of the momentum in the horizontal direction.

$$\Delta p \approx \frac{h}{\Delta y} = \frac{2h}{w}$$

Let's calculate the location of the edge of this central maximum by assuming that the sideways component of the momentum p is equal to Δp.

$$y = vt = \frac{pt}{m}$$

Substituting in our value for p and realizing that $t = L/v$, we have

$$y = \frac{pt}{m} = \frac{1}{m}\frac{2h}{w}\frac{L}{v} = \frac{2hL}{wmv}$$

Plugging in the numerical values, we get our estimate.

$$y = \frac{2hL}{wmv} = \frac{2\left(6.63\times10^{-34}\text{ J}\cdot\text{s}\right)\left(1\text{ m}\right)}{\left(20\times10^{-6}\text{ m}\right)\left(9.11\times10^{-31}\text{ kg}\right)\left(10^{4}\text{ m/s}\right)} \approx 7\text{ mm}$$

Therefore, the width of the central maximum of the diffraction pattern is roughly 14 mm.

We can check to see if this makes sense by looking at the wave aspects of the electrons using the techniques of Section 15.8 with $m = 1$. This yields

$$y_{min} = \frac{\lambda L}{w}$$

Using the de Broglie relationship from the last section, we can insert the wavelength of the electrons.

$$y_{min} = \frac{\lambda L}{w} = \frac{h}{mv}\frac{L}{w} = \frac{hL}{wmv}$$

This agrees with our previous result within a factor of two. We find that the two methods give the same result, although they approach the problem from very different directions.

Example 24.3.2

If the half-life of an excited state in an atom is 10^{-8} s, what is the uncertainty in the energy of the photon emitted when electrons in this state jump to the ground state?

This time we use the uncertainty principle between energy and time.

$$\Delta E\,\Delta t > h$$

Set the two sides of the relationship "approximately equal" to each other and assume that the uncertainty in the time it takes the excited state to decay is equal to the half-life. Then solving for ΔE, we get the following for the uncertainty in the energy of the excited state.

$$\Delta E \simeq \frac{h}{\Delta t} = \frac{6.63\times10^{-34}\text{ J}\cdot\text{s}}{10^{-8}\text{ s}} = 6.63\times10^{-26}\text{ J} \simeq 4\times10^{-7}\text{ eV}$$

Because the ground state is stable, we can safely assume that the uncertainty in its energy is even smaller. This uncertainty in the energy level is less than one millionth of an electron volt. Although the uncertainty in the energy of the emitted photons is very small compared to their average energy, the uncertainty can be observed as a broadening of the spectral line.

Problems

1. What is the wavelength of an electron traveling at 10% of the speed of light?
2. A proton has a speed of 3×10^6 m/s. What is its wavelength?
3. How big is the wavelength of a car (mass = 2000 kg) traveling at 30 m/s compared to the size of an atom (0.1 nm)?
4. What is the wavelength of a bowling ball (mass = 7.27 kg) rolling down an alley at 7 m/s?
5. At what speed would an electron have to travel to have a wavelength equivalent to that of orange light (600 nm)?
6. At what speed would a proton have to travel to have a wavelength of 10 nm, which corresponds to a wavelength of ultraviolet light?
7. What is the wavelength of a proton with a kinetic energy of 1 eV? Why is this different than the value calculated for the 1-eV electron in Example 24.1.1?
8. If a proton has a kinetic energy of 10 eV, what is its wavelength? How does this value compare to that of the 10-eV electron in the practice problem of Example 24.1.1?

*9. What is the kinetic energy of a proton that has a wavelength equal to the diameter of a hydrogen atom (0.1 nm)?

*10. What is the kinetic energy of an electron with a wavelength equal to the diameter of a hydrogen atom (0.1 nm)?

11. Find the minimum velocity of a 1 mg particle confined to a 1-cm long box.

*12. A virus is the smallest thing that can be seen in an electron microscope. Suppose that a virus with a size of 1 nm, and a density equal to that of water (1 g/cm^3) is localized in a one-dimensional box twice its size. What is the minimum speed of the virus?

13. Let's model the nucleus of an atom as a one-dimensional box with a length of 2×10^{-15} m. What is the kinetic energy of a proton confined to this box?
14. If we replace the proton in the previous problem with an electron, what is the kinetic energy of the electron? What does this tell us about the possible existence of an electron in a nucleus?

*15. In order for an electron to be confined in a nucleus, its wavelength must be smaller than the size of the nucleus (say 2×10^{-15} m). Use the uncertainty principle to calculate the kinetic energy of such an electron non-relativistically? What does this say about the possibility of an electron existing in a nucleus?

*16. Obviously a neutron can exist in a nucleus. Using the idea in the previous problem, show that a neutron would have a kinetic energy much smaller than its rest mass energy.

17. If a beam of electrons pass through a vertical slit with a width of 30 μm, what is the uncertainty in the horizontal component of their velocity?
18. If a light is used to locate an electron to an uncertainty of 4 nm, what is the uncertainty in the velocity of the electron?
19. If the velocity of an electron is determined with an uncertainty of 10 m/s, what is the uncertainty in its position?
20. If you can measure the speed of an electron of energy 40 eV to an uncertainty of 0.1%, what is the uncertainty in its position?
21. A beam of electrons with a speed of 2×10^5 m/s passes through a vertical slit (width = 50 μm), forming a diffraction pattern on a screen 75 cm away. Use the uncertainty principle to calculate the width of the central maximum.
22. What happens to the width of the diffraction pattern in the previous problem if a proton beam is used?

23. A beam of electrons passes through a vertical slit of width 80 μm and produces a diffraction pattern whose central maximum has a width of 16 mm. The screen is 1.2 m from the slit. Use the uncertainty principle to estimate the speed of the electrons.

*24. A beam of electrons with energy 2 eV passes through a slit of width 10 μm producing a diffraction pattern on a screen. The width of the central maximum is 1 mm. How far is the screen located from the slit?

25. If an excited state has a half-life of 10^{-16} s, what is the uncertainty in the energy of the state?

26. If the uncertainty in the energy of an excited state is a billionth of an electron volt, what is the half-life of the state?

25 — THE NUCLEUS

25.1 Radioactive Decay

Half the number of nuclei in a radioactive sample decay in a characteristic time known as the half-life $T_{1/2}$. After a second half-life has elapsed, one-half of the remaining nuclei will have decayed. After a third half-life, one-half of those still remaining decay. And so on. This means that one-half of the original sample remains after $T_{1/2}$, one-half of one-half (or one-fourth) remain after $2T_{1/2}$, and one-half of one-fourth (or one-eighth) of the original sample remains after $3T_{1/2}$. And the process theoretically continues forever. This can be written mathematically in the following way.

$$N = N_o\left(\frac{1}{2}\right)^n$$

where N is the number of nuclei remaining, N_o is the number of nuclei present at the beginning of the time period, and n is the number of half-lives that have elapsed.

The activity of a radioactive sample depends on the number of nuclei that decay in a unit of time. Usually this is the number of decays per second. As an example, the activity of 1 g of radium is 3.7×10^{10} decays/s, a unit named the curie (Ci), after Marie Curie. Because the number of decays is equal to the change in the number of nuclei remaining, we can write the activity as $\Delta N/\Delta t$. Changing the N to ΔN in the equation above and dividing both sides by Δt, we get a relationship that shows how the activity of a sample decreases with time.

$$\frac{\Delta N}{\Delta t} = \left(\frac{\Delta N}{\Delta t}\right)_o\left(\frac{1}{2}\right)^n$$

where the first term in parentheses denotes the activity at the start of the time period. Because the activity is proportional to the number of nuclei present at that time, the number of nuclei and the activity change at the same rate.

Example 25.1.1

The half-life of radium is 1620 years. A 1-g sample of radium contains 2.66×10^{21} nuclei. How many of these will remain after 8000 years?

We first need to know the number of half-lives that have elapsed. This is obtained by dividing the time period T by the half-life $T_{1/2}$.

$$n = \frac{T}{T_{1/2}} = \frac{8000 \text{ y}}{1620 \text{ y}} \simeq 5$$

where we have rounded off the value of n to the nearest integer to avoid computing fractional powers of $\frac{1}{2}$. We can now calculate how many nuclei will remain after 5 half-lives.

$$N = N_o\left(\frac{1}{2}\right)^n = N_o\left(\frac{1}{2^5}\right) = \frac{2.66\times10^{21}}{32} = 8.31\times10^{19}$$

Practice: How many radium nuclei will remain after an additional 8000 years?
Answer: 2.6×10^{18}

Example 25.1.2

What is the activity of the 1-g of radium after 8000 years?

Because the original activity is 1 Ci by definition, the activity after 8000 years will be

$$\frac{\Delta N}{\Delta t} = \left(\frac{\Delta N}{\Delta t}\right)_o\left(\frac{1}{2}\right)^n = \frac{1\,\text{Ci}}{32} = 0.0313\,\text{Ci} = 1.16\times10^9\ \text{decays/s}$$

25.2 Radioactive Clocks

The decay of a radioactive sample is so regular and predictable it can be used as a clock. We need only monitor the activity of a sample to know how much time has elapsed. If we can determine how many half-lives n have elapsed, we can calculate the time T from a knowledge of the half-life $T_{1/2}$.

$$T = nT_{1/2}$$

One of the most common radioactive decays used in the dating of organic substances is that of carbon-14. The ratio of C-14 to the stable C-12 has a constant value of 1.3×10^{-12} in the atmosphere. (High-energy particles from space create C-14 via collisions with nuclei to replace the ones that decay.) Although this means that only about one in a trillion carbon atoms is radioactive, this is enough to be easily detected. As an example, a 1-g sample of carbon has about 5×10^{22} atoms, of which some 65 billion are radioactive. The activity of this sample would be about 16 decays/min.

Example 25.2.1

A sample of carbon from a piece of charred wood from a campfire has an activity of 4 decays/min/g. How long ago was the campsite occupied?

Because the activity was 16 decays/min/g when the wood was chopped down for the fire, the activity has been reduced by a factor of 4. Therefore, two half-lives have elapsed, or

$$T' = nT_{1/2} = 2\,(5700\ \text{y}) = 11{,}400\ \text{y}$$

Practice: How old is the campsite if the activity is only 2 decays/min/g?
Answer: 17,100 y

For samples considerably older than 60,000 years, potassium-argon dating is often used. Naturally occurring potassium is 0.01% radioactive potassium-40, which decays spontaneously; 11% decays to calcium-40 via beta decay and 89% decays to inert argon-40 via electron capture. The latter decay has a half-life of 1.28×10^9 years.

When rock is molten, any gases (including argon) can escape. This sets the radioactive clock to zero at the moment the rock solidifies. Grinding a sample of rock and comparing the ratio of argon-40 to potassium-40 yields the date when the rock solidified.

Example 25.2.2

If we grind a piece of volcanic rock and find that it contains three times as many argon-40 atoms as potassium-40 atoms, approximately how long has it been since this rock solidified?

We choose to simplify the problem by ignoring the secondary decay mode to calcium. When the rock solidified, it contained zero argon atoms. Since that time three-quarters of the potassium-40 atoms have decayed, indicating that two half-lives have passed. The rock has been solid for approximately

$$T \approx 2T_{1/2} = 2\,(1.28 \times 10^9 \text{ y}) = 2.56 \times 10^9 \text{ y}$$

Practice: How long has a rock been solid if it contains seven times more argon atoms than potassium atoms?
Answer: 3.84×10^9 y

25.3 Radiation and Matter

The passage of gamma rays through matter is analogous to radioactive decay. The number of gamma rays surviving after a characteristic *half-distance* is one-half of those at the beginning. During the next half-distance, only one-half of the half survive. And so on. Therefore, repeating the logic of Section 25.1, we can write down the number N surviving after n half-distances as

$$N = N_o \left(\frac{1}{2}\right)^n$$

where N_0 is the initial number.

Example 25.3.1

If a burst of 1000 gamma rays with an energy of 5 MeV is incident on a 18.2-cm thick aluminum plate, how many gamma rays exit the other side of the plate?

From Table 25-4 in the text, we find that the half-distance for 5-MeV gamma rays in aluminum is 9.1 cm. Therefore, the plate is two half-distances thick and the number of gamma rays coming out the other side is

$$N = N_o\left(\frac{1}{2}\right)^n = 1000\left(\frac{1}{2^2}\right) = 250$$

Problems

1. What is the activity of 1 kg of radium?
2. How much radium would you need to have an activity of 10^{12} decays/s?
3. How many nuclei from an initial sample of 4×10^{24} radioactive nuclei will remain after 6 half-lives?
4. If you initially have 10^{20} radioactive nuclei, how many will remain after 8 half-lives?
5. Cesium-137 decays via beta decay to barium-137. You find an old sample whose activity was measured 90 years ago to be 8 Ci. You measure its current activity and find it to be 1 Ci. What is the half-life of cesium-137?
6. If the number of radioactive nuclei in a sample is reduced from 1.6×10^{21} to 4×10^{20} in one hour, what is the material's half-life?

*7. Potassium-40 is a naturally occurring radioactive nucleus that is present in our bodies and is the major internal contributor to the background radiation we experience. With a half-life of 1.3×10^9 y, what change in the activity of potassium-40 has occurred since Earth was formed some 4.5 billion years ago?

8. Scandium-46 beta decays to titanium-46 with a half-life of 84 days. If the current activity of a sample is measured at 4 mCi, what was the sample's activity 336 days earlier?

*9. Tritium (the heavy, heavy isotope of hydrogen) decays via beta minus decay with a half-life of 12.3 y. How long must one wait for more than 99% of a sample of tritium to decay?

10. One of the radioactive nuclei produced in nuclear reactors is uranium-239, which undergoes beta minus decay with a half-life of 23.5 min. How long after the reactor is shut down will the number of uranium-239 nuclei be reduced by a factor of 1000?
11. Zirconium-87 decays to yttrium-87 via beta plus decay with a half-life of 1.7 h. If a sample of zirconium-87 has a current activity of 5 μCi, how long ago was its activity 60 μCi?
12. The label on a commercially purchased radioactive sample gives its activity as 16 μCi. The half-life of the sample is 70 days. You measure the sample's activity and find it to be only 0.25 μCi. How old is the sample?
13. The activity of a gram of carbon from a piece of an ancient boat is 240 decays/h. What is the age of the boat?
14. If a gram of carbon from an axe handle has an activity of 1 decay/min, how old is the axe handle?

*15. If the activity of a gram of carbon is 11.3 decays/min, how old is the sample?

16. If 5 g of carbon from the leg bone of a mummified goat has an activity of 10 decays/min, what is the age of the mummy?
17. The half-life of uranium-238 is 4.5 billion years. If its activity has decreased by a factor of 2 since the formation of Earth, how old is Earth?
18. Tritium has a half-life of 12.3 y and can also be used to date organic matter. If the activity of tritium is determined to be 3% of the ex-

pected activity, what is the age of the sample?

19. The half-distance of 15-MeV gamma rays in lead is 1 cm. If 1 billion 15-MeV gamma rays enter a lead block with a thickness of 4 cm, how many will emerge from the other side?

20. A beam with an intensity of 10^6 gamma rays per second hits an aluminum shield with a thickness of 1 m. If the gamma rays have an energy of 10 MeV, what is the intensity of the beam exiting the shield?

21. What thickness of aluminum is required to reduce the intensity of a 10-MeV beam of gamma rays to less than 0.1% of the original intensity?

22. You wish to reduce the intensity of a beam of 1-MeV gamma rays from 5.12×10^7 per second to 2×10^5 per second. What thickness of aluminum is required?

26 — NUCLEAR ENERGY

26.1 Nuclear Binding Energy

In this chapter we will be discussing very small effects and will use more than our customary three significant digits. The experiments, however, are very precise, so we can justify doing this. It is useful to have the energy equivalent of 1 amu when doing calculations with nuclear binding energies. Using Einstein's formula $E = mc^2$, we have

$$E = \left(1.6605\times10^{-27}\ \text{kg}\right)\left(2.9979\times10^{8}\ \text{m/s}\right)^2\left[\frac{1\ \text{MeV}}{1.6022\times10^{-13}\ \text{J}}\right] = 931.44\ \text{MeV}$$

Example 26.1.1

Given that the mass of an iron-56 nucleus is 55.918 4 amu, what is the average binding energy of $^{56}_{26}\text{Fe}$?

Begin by calculating the mass of the 26 protons and 30 neutrons that make up the iron nucleus. Then look up the mass of the iron nucleus and calculate the mass difference.

$$\begin{aligned} 26\,m_\text{p} &= 26\times 1.007\ 28\ \text{amu} = & 26.189\ 28\ \text{amu} \\ 30\,m_\text{n} &= 30\times 1.008\ 67\ \text{amu} = & +\ \underline{30.260\ 10}\ \text{amu} \\ & & 56.449\ 38\ \text{amu} \\ m_\text{Fe} &= & -\ \underline{55.918\ 4}\ \text{amu} \\ & & 0.5310\ \text{amu} \end{aligned}$$

We now convert this mass to its energy equivalent using the conversion factor that we developed earlier.

$$(0.531\ \text{amu})\left[\frac{931.44\ \text{MeV}}{1\ \text{amu}}\right] = 494.6\ \text{MeV}$$

Dividing this energy by 56, the number of nucleons, we get the average binding energy to be 8.83 MeV/nucleon.

Practice: What is the average binding energy of the deuteron if it has a mass of 2.013 55 amu?
Answer: 1.12 MeV/nucleon

Example 26.1.2

Estimate the mass of $^{240}_{94}\text{Pu}$ given that the binding energy for nuclei this size is 7.5 Me/V nucleon.

The total binding energy of the nucleus is

$$(240 \text{ nucleons})\left(7.5\frac{\text{MeV}}{\text{nucleon}}\right) = 1800 \text{ MeV}$$

which can be converted to its mass equivalent.

$$(1800 \text{ MeV})\left[\frac{1 \text{ amu}}{931.44 \text{ MeV}}\right] = 1.93 \text{ amu}$$

We now add up the masses to get the total mass of the plutonium-240.

$$\begin{aligned} 94\, m_p &= 94 \times 1.00728 \text{ amu} = 94.684\,32 \text{ amu} \\ 146\, m_n &= 146 \times 1.00867 \text{ amu} = +\underline{147.265\,82} \text{ amu} \\ & \qquad\qquad 241.950\,14 \text{ amu} \\ \text{binding energy} &= -\ \underline{1.93} \text{ amu} \\ & \qquad\qquad 240.02 \text{ amu} \end{aligned}$$

where we rounded off the answer because we only had 3 significant digits in the binding energy. This answer compares well with the actual mass of 240.00 amu.

26.2 Nuclear Fission

We can use the graph of the binding energy per nucleon (Fig. 26-5 in the text) to obtain an estimate of the energy released in the fission process. Because the fission reaction must conserve the number of nucleons, any change in energy must come from differences in the binding energy.

Example 26.2.1

If the average binding energies for uranium, barium, and krypton are 7.6, 8.4, and 8.5 MeV/nucleon, respectively, how much energy is released in the following fission process?

$$^{1}_{0}\text{n} + ^{235}_{92}\text{U} \rightarrow ^{141}_{56}\text{Ba} + ^{92}_{36}\text{Kr} + 3\,(^{1}_{0}\text{n})$$

The initial binding energy is

$$(235 \text{ nucleons})\left(7.6\frac{\text{MeV}}{\text{nucleon}}\right) = 1786 \text{ MeV}$$

The binding energy of the fission products is

$$(141 \text{ nucleons})\left(8.4\frac{\text{MeV}}{\text{nucleon}}\right) + (92 \text{ nucleons})\left(8.5\frac{\text{MeV}}{\text{nucleon}}\right) = 1966 \text{ MeV}$$

Therefore, 180 MeV are released in the fission reaction. The average energy released in all of the fission reactions for uranium-235 is 208 MeV.

26.3 Nuclear Reactors

In a nuclear reactor using uranium as fuel, the average amount of energy released in each fission reaction is 208 MeV. Knowing this allows us to calculate how many reactions must take place each second to release a given amount of energy. We begin by converting the energy in MeV to the equivalent value in joules.

$$(208 \text{ MeV})\left[\frac{1.6\times10^{-13} \text{ J}}{1 \text{ MeV}}\right] = 3.33\times10^{-11} \text{ J}$$

This means that every watt requires

$$1 \text{ W} = \left(1\frac{\text{J}}{\text{s}}\right)\left(\frac{1 \text{ reaction}}{3.33\times10^{-11} \text{ J}}\right) = 3\times10^{10} \text{ reactions/s}$$

Example 26.3.1

How many fission reactions per second are required to release 20 MW of thermal energy?

$$20\times10^{6} \text{ W} = \left(\frac{3\times10^{10} \text{ reactions/s}}{1 \text{ W}}\right) = 6\times10^{17} \text{ reactions/s}$$

Practice: What is the rate of fission reactions required for a 400-MW reactor?
Answer: 1.2×10^{19} reactions/s

Example 26.3.2

Uranium ore contains 0.7% U-235, most of the rest being U-238. If we could somehow collect a kilogram of pure U-235 and all of it were to fission, how much energy would be released?

Because we know that 235 g of U-235 contains Avogadro's number of atoms, we have a total of

$$\left(6.02\times10^{23}\text{ nuclei}\right)\left(\frac{1000\text{ g}}{235\text{ g}}\right)=2.56\times10^{24}\text{ nuclei}$$

The energy released by these nuclei is

$$\left(2.56\times10^{24}\text{ nuclei}\right)\left(\frac{3.33\times10^{-11}\text{ J}}{1\text{ nucleus}}\right)=8.52\times10^{13}\text{ J}$$

To get a better feeling for how much energy this is, let's convert the energy to kilowatt-hours. A kilowatt-hour is the energy provided by a kilowatt power source during each hour of operation.

$$\left(8.52\times10^{13}\text{ J}\right)\left[\frac{1\text{ kWh}}{3.6\times10^{6}\text{ J}}\right]=2.37\times10^{7}\text{ kWh}$$

Let's assume an efficiency of 35% for converting this thermal energy to electrical energy, and that a typical household uses 12,000 kWh/y. Then we can calculate the number of households that this kilogram could supply.

$$\left(2.37\times10^{7}\text{ kWh}\right)(0.35)\left(\frac{1\text{ household}}{12{,}000\text{ kWh}}\right)\simeq 700\text{ households}$$

26.4 Fusion Reactors

Example 26.4.1

How much energy is released in the following reaction in which two isotopes of hydrogen—deuterium and tritium—fuse to form helium and a neutron?

$${}^{2}_{1}\text{H}+{}^{3}_{1}\text{H}\rightarrow{}^{4}_{2}\text{He}+{}^{1}_{0}\text{n}$$

We can get our answer by taking the difference in the total mass before the reaction and the total mass after the reaction.

$$\begin{aligned} m_d &= 2.013\ 55\text{ amu}\\ m_t &= +3.015\ 50\text{ amu}\\ m_{He} &= -4.001\ 50\text{ amu}\\ m_n &= -\underline{1.008\ 67}\text{ amu}\\ & 0.018\ 88\text{ amu}\end{aligned}$$

Converting this mass to energy, we find that 17.6 MeV of energy are released. This is about 3.5 MeV/nucleon. Therefore, this process is more efficient than the fissioning of uranium.

26.5 Solar Power

The source of the Sun's energy is the fusion of hydrogen to form helium. Although the process takes place in many steps, the net effect can be summarized as follows.

$$4\,{}^{1}_{1}\text{H} \rightarrow {}^{4}_{2}\text{He} + 2\,\beta^{+} + 2\,\nu + \text{ energy}$$

Including the energy of the neutrinos ν and the photons due to the annihilation of the two positrons, the total energy released is 26.7 MeV.

Example 26.5.1

How much energy would be released by the complete fusion of 1 kg of hydrogen to form helium?

We know that 1 gram of hydrogen contains Avogadro's number of hydrogen nuclei, so 1 kg will contain 1000 times as many, or 6.02×10^{26} nuclei. Because each four hydrogen nuclei will release 26.7 MeV, the total energy released is

$$\left(6.02\times10^{26}\text{ nuclei}\right)\left(\frac{26.7\text{ MeV}}{4\text{ nuclei}}\right) = 4.02\times10^{27}\text{ MeV} = 6.43\times10^{14}\text{ J}$$

Therefore, we see that 1 kg of hydrogen releases about 7.5 times as much energy as 1 kg of uranium (see Example 26.3.2).

Problems

1. An electon is accelerated through a potential difference of 5×10^9 V. For energies much greater than the rest-mass energy ($E_o = mc^2 =$ 0.511 MeV for an electron), a relativistic treatment yields a momentum of E/c, where c is the speed of light. What wavelength does this electron have?
2. Use the information in Problem 1 to find the energy of a relativistic electron with a wavelength of 5×10^{-16} m.
3. Given that the mass of the helium nucleus is 0.0304 amu less than its constituents, verify that the binding energy for the helium nucleus is equal to 28.3 MeV.
4. If the Pb-206 nucleus has a mass 1.7425 amu less than the total mass of its neutrons and protons, what is its average binding energy in MeV/nucleon?
5. Calculate the average binding energy per nucleon of the C-12 nucleus given that its mass is 11.9967 amu. (The mass of the neutral atom is 12.0000 amu. The mass of the 6 electrons has been subtracted from this and the binding energy of the electrons to the nucleus has been neglected.)
6. Given that the mass of the neutral Pu-239 atom is 239.052 157 amu, calculate the average binding energy per nucleon of the Pu-239 nucleus. (*Hint*: Begin by finding the mass of the bare Pu-239 nucleus.)
7. The average binding energy is about 8 MeV /nucleon when the total number of nucleons

is around 170. What is the estimated mass of the Yb-170 nucleus?

8. The average binding energy is about 8.4 MeV /nucleon when the total number of nucleons is around 70. What is the estimated mass of the Br-78 nucleus?

9. Use the atomic masses given below to calculate the energy released when U-236 fissions to produce Cs-141 and Rb-93.

n	1.008 67
U-236	236.045 56
Cs-141	140.919 63
Rb-93	92.921 57

10. If a uranium nucleus were to split evenly into two nuclei with the same number of nucleons (a rare occurrence) while releasing 2 neutrons, how much energy would be released?

11. A fission reactor is being designed with a thermal power rating of 500 MW. How many fission reactions must take place each second?

12. How many fission reactions must take place each second in a nuclear reactor with a thermal rating of 4 GW?

13. How many grams of U-235 would be needed each hour in a power plant with a thermal rating of 500 MW?

14. The total world uses energy at a rate of 7×10^{12} W. How many grams of U-235 would be required each hour to produce this power?

15. Assume that all of the U-235 in natural uranium ore could be converted to electrical energy with an efficiency of 35%. How many kilograms of natural uranium ore would be required to power an average city of 15,000 households for 1 year?

16. The total world uses energy at a rate of 7×10^{12} W. How many kilograms of hydrogen would be required each second to produce this power?

17. Given that the mass of ${}^{3}_{2}\text{He}$ is 3.014 93 and using the masses in Example 26.4.1, how much energy is released in the following fusion reaction?

$$ {}^{2}_{1}\text{H} + {}^{2}_{1}\text{H} \rightarrow {}^{3}_{2}\text{He} + {}^{1}_{0}\text{n} $$

18. Given that the mass of ${}^{3}_{2}\text{He}$ is 3.014 93 and using the masses in Example 26.4.1, how much energy is released in the following fusion reaction?

$$ {}^{3}_{2}\text{He} + {}^{3}_{2}\text{He} \rightarrow {}^{4}_{2}\text{He} + {}^{1}_{1}\text{H} + {}^{1}_{1}\text{H} $$

*19. Given that the power output of the Sun is 3.83×10^{26} W, how many hydrogen nuclei must be converted to helium each second?

20. Given that the present energy radiated by the Sun each second is about 3.83×10^{26} W, what is the decrease in the mass of the Sun in 1 h?

27 — ELEMENTARY PARTICLES

27.1 Antimatter

When a positron—the antiparticle of the electron—is slowed down in matter, it is usually captured by an electron to form an "atom" called *positronium*. The electron and positron orbit each other about their common center of mass. Within a microsecond, the two annihilate each other to produce two or more photons. If the positronium was at rest, the total momentum before the annihilation was zero. Therefore, it must be zero afterward. If there are only two photons produced, they must leave with equal and opposite momenta. This means that each photon carries off one-half of the total energy. Because the positron and the electron have the same mass, each photon must have an energy equal to the rest-mass energy of the electron. (We are ignoring the orbital kinetic energy and the electric potential energy, which are very small compared to the rest-mass energy.)

Example 27.1.1

What is the energy of one of the photons emitted in the annihilation of a positron and an electron?

$$E = m_e c^2 = \left(9.11\times10^{-31}\text{ kg}\right)\left(3\times10^{8}\text{ m/s}\right)^2 = 8.2\times10^{-14}\text{ J}$$

Because a joule is a very large energy unit on the atomic and subatomic scales, it is common practice to convert this to electron volts.

$$E = \left(8.2\times10^{-14}\text{ J}\right)\left[\frac{1\text{ eV}}{1.6\times10^{-19}\text{ J}}\right] = 5.12\times10^{5}\text{ eV} = 0.512\text{ MeV}$$

Therefore, each photon has about one-half a million electron volts of energy.

Let's calculate the frequency of these photons to determine their place in the electromagnetic spectrum.

$$f = \frac{E}{h} = \frac{8.2\times10^{-14}\text{ J}}{6.63\times10^{-34}\text{ J}\cdot\text{s}} = 1.24\times10^{20}\text{ Hz}$$

This frequency is in the ranges of X rays and gamma rays. They are usually called gamma rays.

Practice: What is the energy (in MeV) of the photons emitted in the two-photon annihilation of a proton and an antiproton?
Answer: 938 MeV

27.2 Exchange Particles

We can use the Heisenberg uncertainty principle to obtain an estimate of the range of the strong force. As discussed in the text, all forces are due to the exchange of particles. For instance, the electromagnetic force is due to the exchange of photons. The force between two nucleons can be considered to be due to the exchange of particles such as the pion.

Consider the interaction between a neutron and a proton at rest. Suppose that the neutron spontaneously emits a pion that is absorbed by the proton. According to classical physics, there is a problem. A neutron at rest has a total energy equal to its rest-mass energy $m_n c^2$, where m_n is the mass of the neutron. Even if the pion is emitted with no momentum, the total energy of the neutron and pion after the emission is $m_n c^2 + m_\pi c^2$, where m_π is the mass of the pion. This is a violation of the conservation of energy by $m_\pi c^2$. In general, the violation will be larger than this because the pion and neutron will each have some kinetic energy.

Although such violations of energy are forbidden in classical physics, they are allowed in quantum physics as long as the violations do not last too long. In practice, the violation is corrected when the proton absorbs the pion on the other end of the process. But how long is too long? The violation lasts too long if the uncertainty principle doesn't prevent us from measuring it. In the limiting case, the uncertainty principle involving energy and time states

$$\Delta E\, \Delta t \approx h$$

where the symbol $\approx$ means "approximately equal to." Therefore, the maximum time Δt that a violation can last depends on the value of ΔE.

We can get an upper estimate of the distance the exchange particle can travel by knowing that it cannot exceed the speed of light. Therefore, the exchange particle cannot travel any farther than

$$d_{max} < c\Delta t \approx \frac{ch}{\Delta E}$$

Example 27.2.1

What is the maximum range of the strong force if it is due to the exchange of pions?

Let's begin by calculating the maximum time the violation can last for the minimum violation of creating the pion at rest. Any other violation would have to last for less time.

$$\Delta t \approx \frac{h}{\Delta E} = \frac{6.63\times10^{-34}\ \text{J}\cdot\text{s}}{(140\ \text{MeV})}\left[\frac{1\ \text{MeV}}{1.6\times10^{-13}\ \text{J}}\right] \approx 3\times10^{-23}\ \text{s}$$

where 140 MeV is the rest-mass energy of the pion. Although this pion was assumed to be at rest, we can still get a maximum range by assuming that it cannot travel faster than light. In a more realistic calculation, the time would be a factor of 10 or more smaller than our number. This is not terribly important as we are only trying to get an idea of the maximum range.

$$d_{max} < c\Delta t = \left(3\times10^{8}\ \text{m/s}\right)\left(3\times10^{-23}\ \text{s}\right) = 9\times10^{-15}\ \text{m} = 9\ \text{fm}$$

Notice that this result is in rough agreement with the range of 3 fermis stated in the text. It is good that the number is larger, as we were calculating a number that we knew was too big.

Practice: If part of the interaction is due to the exchange of kaons (mass = 494 MeV/c^2), what is the maximum range of this part of the interaction?
Answer: 3 fm

Example 27.2.2

The maximum range of the weak force has been found to be roughly 10^{-17} m (a few percent of the diameter of a nucleus). Use this fact to estimate the mass of the exchange particles that mediate the weak force.

Again we can use the Heisenberg uncertainty principle to estimate

$$\Delta E \approx \frac{h}{\Delta t} < \frac{hc}{d_{\text{max}}} = \frac{\left(6.63\times10^{-34}\ \text{J}\cdot\text{s}\right)\left(3\times10^{8}\ \text{m/s}\right)}{10^{-17}\ \text{m}}\left[\frac{1\ \text{eV}}{1.6\times10^{-19}\ \text{J}}\right]$$
$$= 1.24\times10^{11}\ \text{eV} \approx 100\ \text{GeV}$$

This rest energy is about two orders of magnitude larger than the rest energy of a proton.

Because photons have zero rest-mass and travel at the speed of light, we can use this relationship to see why the electrostatic force has an infinite range and why it gets weaker with increasing distance. If the energy of the exchanged photon is small, its effect on the particles is small. This makes sense because the photon has a small momentum. According to our relationship, a small ΔE means a large range. Therefore, the force is weak at long range. On the other hand, a photon with a larger energy will cause a stronger interaction. However, it has a larger ΔE and, consequently, a shorter range.

Example 27.2.3

What is the maximum range of a 1-eV photon emitted by a proton at rest?

$$d_{max} < \frac{ch}{\Delta E} = \frac{\left(3\times10^{8}\ \text{m/s}\right)\left(6.63\times10^{-34}\ \text{J}\cdot\text{s}\right)}{(1\ \text{eV})}\left[\frac{1\ \text{eV}}{1.6\times10^{-19}\ \text{J}}\right] \approx 10^{-6}\ \text{m}$$

Problems

1. The negative kaon is the antiparticle of the positive kaon (and vice versa). If these two kaons annihilate to form two photons, what is the energy of each? The mass of each kaon is 8.78×10^{-28} kg.

2. If a neutron and an antineutron annihilate at rest to form two photons, what is the energy (in MeV) of each photon?
3. What is the frequency of the photon in problem 1?
4. What is the frequency of the photon in problem 2?
5. What is the maximum range of a 1-MeV photon emitted by a particle at rest?
6. If a particle at rest emits a 1-keV photon, what maximum range will it have?
7. What is the maximum possible energy of a photon of range 1 nm?
8. What is the maximum possible energy of a photon of range 1 m?
9. What energy exchange photon would have the same maximum range as that for the pion in Example 27.2.1?
10. The maximum range of the kaon can be calculated to be about 3 fm. What energy exchange photon has the same maximum range?
11. What is the maximum range of the interaction due to the exchange of eta prime mesons with a rest-mass energy 958 MeV?
12. The rest-mass energy of the W intermediate vector boson is 80.6 GeV. What is the range of this part of the weak force?
13. If the range of the part of the weak force mediated by the Z intermediate vector boson is 5×10^{-18} m and its rest-mass energy is 91.2 GeV, what is the average speed of the Z boson?
14. If the rest-mass energy of the Z intermediate vector boson is 91.2 GeV, what is the range of this part of the weak force?

ANSWERS TO ODD-NUMBERED PROBLEMS

Chapter 1

1. a) 3.34 b) 38,600 c) 0.667 d) 0.001 23
3. a) 7.17 b) 59,800
5. a) 4.77×10^{-3} b) 2.05×10^{2}
7. 509 mph
9. 1.95 m^3
11. 6.34×10^4 in.
13. 183 cm
15. 5.2×10^6 mm
17. 113 km/h
19. 115 km/h
21. 10,000 cm^2
23. 10^9 breaths/lifetime
25. 2.8×10^8 gal/day

Chapter 2

1. 15.4 km/h
3. 45 mph
5. 22.7 mph
7. 11.7 mph
9. 1140 km
11. 22.4 days
13. ≈ 1.9 m/s
15. ≈ -2.3 m/s^2
17. 36 km/h·s
19. a) 43.1 mph/s b) 153 mph
21. −90 km/h·min (south)
23. 900 m/s^2 (up)
25. −1.6 m/s^2 (down)
27.

t(s)	v(m/s)	d(m)
0	0	0
1	9.8	4.9
2	18.6	19.6
3	29.4	44.1
4	39.2	78.4
5	49.0	122.5

29. a) 0 b) 3 s c) 3 s d) 44.1 m e) 29.4 m/s
31. 250 m; 300 m
33. 51.6 m
35. 88 ft; 176 ft

Chapter 3

1. 70.7 N northwest
3. 72.1 N at 56.3° above the horizontal to the left
5. 7200 N east
7. 9.8 m/s^2 to the right
9. 8 kg
11. 3; red
13. 2 s
15. a) greater b) up c) yes
17. $W = 44.4$ N; $m = 12$ kg
19. 935 N
21. 61.2 N
27. 432 N
29. 650 N
31. 196 N
33. 0.51

Chapter 4

1. a) 3 m/s west; b) 3 m/s east; c) 9 m/s east
3. 141 mph southwest
5. 6.71 m/s at 26.6° north of west
7. a) 11 m/s west; b) 2 m/s west
9. 0.71 m/s^2
11. 368 N
13. 15 m/s; 2.2 m/s (upward)
15. a) 2 s b) 40 m
17. 49.6 m/s
19. 30 m

Chapter 5

1. 5.93×10^{-3} m/s^2; same
3. 6.57×10^{-6} m/s^2
5. 5.58×10^{22} N
7. 0.00544%
9. The object with the smaller acceleration has 9 times the mass
11. 3.44×10^{-5} N; 3.22×10^{-5} N
13. 0.044 4 times as large

15. Jupiter exerts 11.8 times as much force
17. 6.46×10^{23} kg
19. 3.69 m/s^2
21. 1.48 m/s^2
23. 0.886 g
25. 5.59×10^3 m/s
27. 238 min
29. 2.42×10^8 km
31. 5.93×10^{24} kg
33. $g/4$ = 2.46 N/kg

Chapter 6

1. −20,000 kg·m/s (backward)
3. −14.5 kg·m/s (backward)
5. 40 s
7. 8400 N; 12.2 times as much
9. 3.06 s
11. 261 N
13. 0.15 m/s
15. 4.76 m/s
17. −1 m/s (left)
19. 1 m/s (right)
21. 22.7 m/s
23. 21.4 m/s

Chapter 7

1. 400 J
3. 28.3 m/s
5. Momentum yes; kinetic energy no
7. $0.6\left[\frac{1}{2}mv_o^2\right]$
9. 6 J
11. 2000 N (opposite motion)
13. 0.48 J
15. 107 ft
17. 300 J
19. 10 cm
21. 7840 J
25. 588 J
27. 13.9 m/s
29. 3.13 m/s
31. 81.6 m
33. 1420 W
35. Converted to gravitational potential energy
37. N·m/s = J/s

Chapter 8

1. 1.25 m
3. 45 kg
5. 13 cm
7. 10 cm from center
9. 3.77 rad/s^2
11. 16.8 rev/s = 106 rad/s
13. 4.5 rev
15. 4 rad/s^2
17. 1.28 kg·m^2
19. 3220 km/h
21. 2.36 kg·m^2/s
23. 0.667 rev/s
25. 30 rev/min

Chapter 9

1. 10 m/s
3. 58.3 m/s
5. 301 N
7. 199 N
9. 10.2 m/s^2
13. 0.408 m
15. 7.4 m/s^2 = 0.755 g
17. 1.96 m
19. 9.23×10^{-11} m/s^2
21. 1.9×10^{-10} m/s^2 = 1.94×10^{-11} g

Chapter 10

1. 50 s; 25 m
3. 3.43 h; 0.1 h
5. 3.2
7. 0.107 ms
9. 0.995 c
11. 0.929 c
13. 0.991 c
15. 1.22×10^{12} m
17. 0.866 c
19. a) 1.03×10^{-18} N·s b) 5.63×10^{-22} N·s
21. 1.03×10^{-12} N
23. 1.51×10^{-10} J
25. 0.968 c
27. 1.38
29. 11.5° from the vertical
31. 2.18×10^{-14} m

33. 3.27×10^6 m/s^2

Chapter 11

1. 40 L
3. 200 L
5. 10^{-2} cm
7. 1000 m^2; less than half a football field
9. 1.25×10^{17} atoms
11. 7.84×10^4 Pa
13. 494 m/s; same if temperatures are the same
15. –40°C; –17.8°C; 20°C
17. –40°F; 68°F; 98.6°F
19. 60°C; 140°F
21. 559°F
23. –273°C; 37°C
25. 545 m/s
27. 2.55 L
29. $V_f = V_i /3$
31. 300 K = 27°C
33. 287 balloons

Chapter 12

1. 926 kg/m^3
3. 623 kg/m^3 = 0.623 g/cm^3; float
5. 0.965 g
7. 73.5 cm^3
9. 163 N/m
11. 0.0368 m = 3.68 cm
13. 0.392 m; 50 N/m
15. 1.18×10^5 N/m^2
17. 3.79 m
19. 1.96×10^5 N/m^2 (above atm)
21. from 5 m to the surface
23. 9.29 N
25. 3.06×10^{-5} m^3 = 30.6 cm^3
27. 1.62 kg
29. 42.5% above the surface

Chapter 13

1. –90 J
3. 120 m
5. 4.41 J/cal; thermal losses
7. 21,400 m
9. 2×10^4 cal
11. 4600 cal
13. 0.214 cal/g·°C
15. 31.2°C
17. 20.2°C
19. 58.8 g
21. 379 cal/°C
23. 39.9 kcal
25. 539 cal compared to 80 cal
27. 79.8 g
29. 1650 W
31. 1450 W
33. 9570 nm
35. 500 nm
37. 7.14 cm
39. 19.991 2 cm
39. 1.0036 L

Chapter 14

1. 0.25 = 25%
3. 0.251 = 25.1%
5. 0.556 = 55.6%
7. 150°C
9. For nuclear: Q_{in} = 3.13 J; Q_{out} = 2.13 J; For coal: Q_{in} = 2.63 J; Q_{out} = 1.63 J; the nuclear wastes 30.7% more energy
11. 6 J
13. 13.7
15. 0.6
17. 6
19. $1/12^2$ = 0.694%
21. 29*k*

Chapter 15

1. T = 10 s; f = 0.1 Hz
3. 6.13 cm
5. 79 N/m
7. 243 kg
9. 1.28 J
11. 6.95 s
13. 3.97 m
15. 3.71 m/s^2
17. 7.6 s
19. 98 N/m; 0.9 s
21. 3 m
23. 40 m/s

25. (4 m)/n with n an odd integer
27. 85.8 Hz
29. 0, ±12.5 cm, ±25 cm, ...
31. 3.43 m

Chapter 16

1. 49 m/s
3. 2.72 s
5. 85.8 m
7. 10^{-3} W/m^2
9. 30 dB
11. 417 Hz
13. $\lambda_1 = 1.32$ m; $\lambda_2 = 0.66$ m; $\lambda_3 = 0.44$ m; $f_1 = 500$ Hz; $f_2 = 1000$ Hz; $f_3 = 1500$ Hz
15. 172 Hz
17. 85.8 Hz
19. 0.3 m
21. 80 cm
23. 421 Hz or 427 Hz
25. 736 Hz
27. 94 Hz
29. 486 Hz; 384 Hz
31. 19.5 m/s

Chapter 17

1. 0.8 m
3. 2 m
5. 237 cm
7. Real; 60 cm in front; –1
9. 7.5 cm
11. 20 cm in front; 4 cm; inverted
13. at infinity
15. –7.5 cm
17. $1/s' = 1/(-f) - 1/s < 0$ ==> always virtual
19. 51.8 y
21. 169,900 y

Chapter 18

1. 2.25 m below the surface
3. 4 m
5. 300 cm on far side; 60 cm
7. 30 cm on near side
9. 10.2 cm
11. 3.33 cm on the near side
13. Virtual; 12 cm on near the side; 0.6
15. 15 cm; 30 cm
17. 2.75 mm
19. 1 cm
21. 59.9 diopters
23. –25 cm
25. 2.33 cm

Chapter 19

1. 2.05×10^8 m/s
3. 1.43
5. 790
7. 226 nm
9. 515 nm
11. 21.8 mm
13. 0.0269 mm
15. 2.94 cm
17. 520 nm
19. 110 nm
21. $(m + \frac{1}{2})$(173 nm) with m an integer
23. m(217 nm) with m a non-zero integer

Chapter 20

1. 1.6×10^{-4} C
3. 2.12×10^{-6} N
5. 2.3×10^{-28} N repulsive
7. 1.72×10^{10} N
9. 3.51×10^{21} m/s^2
11. 5.71×10^{13} C
13. 9.38×10^3 N/C south
15. 0.734 m
17. 1.69×10^5 N/C
19. 2.53×10^{28} m/s^2
21. 80 J
23. 48 J; same
25. 45 J; –45 J
27. 33.3 cm
31. 375 V

Chapter 21

1. 58.8 m
3. 4 Ω
5. 0.99 Ω; therefore consistent

7. 18 Ω
9. 2 Ω; 4 Ω; 9 Ω; 18 Ω
11. 6 Ω in series
13. 2 V; 4 V
15. 1.5 A; 0.5 A
17. 1.33 Ω; 4 V; 2 A
19. 5 V
21. A = E > D > B = C
23. Circuit 2
25. 5 = 6 > 1 = 4 > 2 = 3
27.

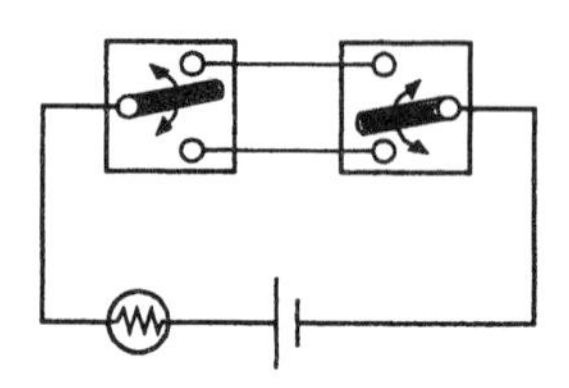

29. 1200 W
31. 26.7 Wh
33. 12¢
35. 24 bulbs
37. 48 W; twice as much
39. 10-W bulb; it has more resistance

Chapter 22

1. 5.12×10^{-14} N
3. 8.33×10^{-5} C
5. 8.54×10^{-4} m
7. 6.83×10^{-5} T
9. 4.79×10^{6} m/s; 1.31×10^{-6} s
11. 0.126 A
13. 50 m/s
15. 1.64 J
17. 150,000 turns
19. 12 V
21. 3.75×10^{18} Hz
23. 1.05×10^{10} Hz
25. 2.98 m
27. 5 m to 5.56 m

Chapter 23

1. 6.41×10^{-26} J
3. 1.51×10^{33} Hz; no
5. 1.24×10^{-6} eV
7. 3.59 eV
9. 1.14×10^{15} Hz
11. 3.44 eV
13. 2.12×10^{-34} J.s
15. 2.12×10^{-10} m
17. –3.4 eV
19. 3.4 eV
21. 3.09×10^{15} Hz; ultraviolet
23. 3.28×10^{15} Hz
25. 435 nm

Chapter 24

1. 2.43×10^{-11} m
3. 1.11×10^{-28}
5. 1.21×10^{3} m/s
7. 2.87×10^{-11} m; different masses
9. 0.0823 eV
11. 3.31×10^{-26} m/s
13. 5.14×10^{7} eV
15. 378 GeV; not possible
17. 48.5 m/s
19. 7.28×10^{-5} m
21. 0.218 mm
23. 2.73×10^{3} m/s
25. 41.4 eV

Chapter 25

1. 3.7×10^{13} decays/s
3. 6.25×10^{22} nuclei
5. 30 y
7. 0.0909
9. Approx. 7 half-lives = 86 y, or more exactly 6.64 half-lives = 81.7 y
11. 6.8 h
13. 11,500 y
15. 2870 y
17. 4.5 billion y
19. 6.25×10^{7}
21. 111 cm

Chapter 26

1. 2.49×10^{-16} m
3. 28.3 MeV
5. 7.68 MeV/nucleon

7. 169.92 amu
9. 174 MeV
11. 1.5×10^{19} reactions/s
13. 21.1 g/h
15. 3060 kg
17. 3.26 MeV
19. 3.59×10^{38} nuclei/s

Chapter 27

1. 494 MeV
3. 1.19×10^{23} Hz
5. 1.24×10^{-12} m
7. 1.24 keV
9. 138 MeV
11. 1.3 fm
13. 0.367 *c*

Appendix

Physical Constants and Data

Acceleration due to Gravity	$g = 9.80$ m/s^2
Atomic Mass Unit	amu = 1.66×10^{-27} kg
Avogadro's Number	$N_A = 6.02 \times 10^{23}$ particles/g-mol
Bohr Radius	$r_1 = 5.29 \times 10^{-11}$ m
Density of Water	$D_w = 1.00 \times 10^3$ kg/m^3
Electron Volt	eV = 1.60×10^{-19} J
Elementary Charge	$e = 1.60 \times 10^{-19}$ C
Coulomb's Constant	$k = 8.99 \times 10^9$ N·m^2/C^2
Gravitational Constant	$G = 6.67 \times 10^{-11}$ N·m^2/kg^2
Mass of Electron	$m_e = 9.11 \times 10^{-31}$ kg = 0.000 549 amu
Mass of Proton	$m_p = 1.673 \times 10^{-27}$ kg = 1.007 276 amu
Mass of Neutron	$m_n = 1.675 \times 10^{-27}$ kg = 1.008 665 amu
Planck's Constant	$h = 6.63 \times 10^{-34}$ J·s
Speed of Light	$c = 3.00 \times 10^8$ m/s
Speed of Sound (20°C, 1 atm)	$v_s = 343$ m/s
Standard Atmospheric Pressure	$P_{atm} = 1.01 \times 10^5$ Pa

Geometry

Pi	$\pi = 3.14159$
Circumference of Circle	$C = 2\pi r$
Area of Circle	$A = \pi r^2$
Volume of Cylinder	$V = \pi r^2 L$
Surface Area of Sphere	$A = 4\pi r^2$
Volume of Sphere	$V = \frac{4}{3}\pi r^3$

Standard Abbreviations

A	ampere
amu	atomic mass unit
atm	atmosphere
Btu	British thermal unit
C	coulomb
^{o}C	degree Celsius
cal	calorie
Ci	curie
eV	electron volt
^{o}F	degree Fahrenheit
ft	foot
g	gram
h	hour
hp	horsepower
Hz	Hertz
in.	inch
J	joule
K	kelvin
kg	kilogram
lb	pound
m	meter
min	minute
mph	mile per hour
N	newton
Pa	pascal
psi	pound per square inch
rad	radian
rev	revolution
s	second
T	tesla
V	volt
W	watt
Ω	ohm

Prefixes for Powers of Ten

10^{-2}	centi	c
10^{-3}	milli	m
10^{-6}	micro	μ
10^{-9}	nano	n
10^{-12}	pico	p
10^{-15}	femto	f
10^{3}	kilo	k
10^{6}	mega	M
10^{9}	giga	G
10^{12}	tera	T

Conversion Factors

Time

$1 \text{ y} = 3.16 \times 10^7 \text{ s}$
1 day = 86,400 s
1 h = 3600 s

Length

1 in. = 2.54 cm
1 m = 39.37 in. = 3.281 ft
1 ft = 0.3048 m
1 km = 0.621 mile
1 mile = 1.609 km
$1 \text{ LY} = 9.461 \times 10^{15} \text{ m}$

Area

$1 \text{ m}^2 = 10.76 \text{ ft}^2$
$1 \text{ ft}^2 = 0.0929 \text{ m}^2$
$1 \text{ cm}^2 = 0.1550 \text{ in.}^2$
$1 \text{ in.}^2 = 6.452 \text{ cm}^2$

Volume

$1 \text{ m}^3 = 35.32 \text{ ft}^3$
$1 \text{ ft}^3 = 0.02832 \text{ m}^3$
$1 \text{ liter} = 1000 \text{ cm}^3 = 10^{-3} \text{ m}^3$
1 liter = 1.0576 quart
1 quart = 0.9455 liter

Mass

1 ton (metric) = 1000 kg
$1 \text{ amu} = 1.66 \times 10^{-27} \text{ kg}$
1 kg weighs 2.2 lb
454 g weighs 1 lb

Force

1 N = 0.2248 lb
1 lb = 4.448 N

Speed

1 mile/h = 1.609 km/h
1 km/h = 0.6215 mile/h
1 m/s = 3.281 ft/s = 2.237 mile/h
1 mph = 0.447 m/s
1 ft/s = 0.3048 m/s = 0.6818 mile/h

Acceleration

$1 \text{ m/s}^2 = 3.281 \text{ ft/s}^2$
$1 \text{ ft/s}^2 = 0.3048 \text{ m/s}^2$

Energy

1 J = 0.738 ft·lb = 0.2389 cal
1 cal = 4.186 J
1 Btu = 252 cal = 1054 J
$1 \text{ eV} = 1.6 \times 10^{-19} \text{ J}$
$1 \text{ J} = 6.241 \times 10^{18} \text{ eV}$
$1 \text{ kWh} = 3.6 \times 10^6 \text{ J}$
931.44 MeV from mass of 1 amu

Power

1 W = 0.738 ft·lb/s
1 hp = 550 ft·lb/s = 0.746 kW
1 Btu/h = 0.293 W

Pressure

$1 \text{ atm} = 1.013 \times 10^5 \text{ Pa}$
1 atm = 14.7 psi = 76 cm Hg

The Greek Alphabet

Alpha	α	A		Nu	ν	N
Beta	β	B		Xi	ξ	Ξ
Gamma	γ	Γ		Omicron	ο	O
Delta	δ	Δ		Pi	π	Π
Epsilon	ε	E		Rho	ρ	P
Zeta	ζ	Z		Sigma	σ	Σ
Eta	η	H		Tau	τ	T
Theta	θ	Θ		Upsilon	υ	Y
Iota	ι	I		Phi	φ	Φ
Kappa	κ	K		Chi	χ	X
Lambda	λ	Λ		Psi	ψ	Ψ
Mu	μ	M		Omega	ω	Ω

Solar System Data

Object	Mass (kg)	Radius (m)	Period (s)	Orbit (m)
Mercury	3.27×10^{23}	2.44×10^{6}	7.60×10^{6}	5.79×10^{10}
Venus	4.90×10^{24}	6.05×10^{6}	1.94×10^{7}	1.08×10^{11}
Earth	5.98×10^{24}	6.37×10^{6}	3.16×10^{7}	1.50×10^{11}
Mars	6.40×10^{23}	3.40×10^{6}	5.94×10^{7}	2.28×10^{11}
Jupiter	1.90×10^{27}	7.14×10^{7}	3.74×10^{8}	7.78×10^{11}
Saturn	5.64×10^{26}	6.00×10^{7}	9.30×10^{8}	1.43×10^{12}
Uranus	8.73×10^{25}	2.62×10^{7}	2.65×10^{9}	2.87×10^{12}
Neptune	1.03×10^{26}	2.52×10^{7}	5.20×10^{9}	4.50×10^{12}
Pluto	1.5×10^{22}	1.1×10^{6}	7.85×10^{9}	5.91×10^{12}
Moon	7.36×10^{22}	1.74×10^{6}		
Sun	1.99×10^{30}	6.96×10^{8}		

Earth-Moon Distance $R_{em} = 3.84 \times 10^{8}$ m